SpringerBriefs in Neuroscie

For further volumes:
http://www.springer.com/series/8878

Steven R. King

Neurosteroids and the Nervous System

Steven R. King
Department of Cell Biology
and Biochemistry
Texas Tech University Health
Sciences Center
Lubbock, TX, USA

ISSN 2191-558X ISSN 2191-5598 (electronic)
ISBN 978-1-4614-5558-5 ISBN 978-1-4614-5559-2 (eBook)
DOI 10.1007/978-1-4614-5559-2
Springer New York Heidelberg Dordrecht London

Library of Congress Control Number: 2012948135

© Springer Science+Business Media, LLC 2013
This work is subject to copyright. All rights are reserved by the Publisher, whether the whole or part of the material is concerned, specifically the rights of translation, reprinting, reuse of illustrations, recitation, broadcasting, reproduction on microfilms or in any other physical way, and transmission or information storage and retrieval, electronic adaptation, computer software, or by similar or dissimilar methodology now known or hereafter developed. Exempted from this legal reservation are brief excerpts in connection with reviews or scholarly analysis or material supplied specifically for the purpose of being entered and executed on a computer system, for exclusive use by the purchaser of the work. Duplication of this publication or parts thereof is permitted only under the provisions of the Copyright Law of the Publisher's location, in its current version, and permission for use must always be obtained from Springer. Permissions for use may be obtained through RightsLink at the Copyright Clearance Center. Violations are liable to prosecution under the respective Copyright Law.
The use of general descriptive names, registered names, trademarks, service marks, etc. in this publication does not imply, even in the absence of a specific statement, that such names are exempt from the relevant protective laws and regulations and therefore free for general use.
While the advice and information in this book are believed to be true and accurate at the date of publication, neither the authors nor the editors nor the publisher can accept any legal responsibility for any errors or omissions that may be made. The publisher makes no warranty, express or implied, with respect to the material contained herein.

Printed on acid-free paper

Springer is part of Springer Science+Business Media (www.springer.com)

Contents

Neurosteroids and the Nervous System

1 Introduction

Steroids from the adrenal cortex and the gonads profoundly affect the central and peripheral nervous systems (CNS and PNS). Such "neuroactive steroids" act directly on targets in the nervous system or following conversion to other metabolites. These steroids have permanent organizational effects on brain development and activational effects in regulating existing neural circuitry, establishing and maintaining new synaptic connections, and mediating plastic structural changes in the adult brain.

The nervous system itself synthesizes steroids de novo (neurosteroids) [1, 2]. The steroids made are not unique to the CNS and PNS, being also represented in the serum. They include progesterone, dehydroepiandrosterone (DHEA), 17β-estradiol (estradiol) and 5α-reduced steroids, 5α-dihydroprogesterone (DHP), 3α,5α-tetrahydrodeoxycorticosterone (THDOC), the androgen 3α-androstanediol, and 3α,5α-tetrahydroprogesterone (allopregnanolone/THP) along with its isomer, 3α,5β-tetrahydroprogesterone (pregnanolone) [3]. These are the most commonly studied neurosteroids and are the focus of this review. Proof they are synthesized within the CNS comes from several lines of evidence. The most striking is that allopregnanolone levels in the CNS only decline 30% 15 day after the removal of the adrenals and testes despite the precipitous drop in serum steroid [4].

Neurosteroids exist as free steroid or conjugates [1, 2, 5–7]. Concentrations in the brain can be much greater than the bloodstream, with levels of pregnenolone 5, dihydrotestosterone (DHT) 10, and allopregnanolone 50-fold higher than plasma and DHEA-S originally estimated to be nearly 20-fold higher [1, 2, 4, 8]. Using a variation on liquid chromatography with tandem mass spectrometry, one group finds the concentration of testosterone to be 17 nM in the adult male rat hippocampus,

S.R. King, *Neurosteroids and the Nervous System*, SpringerBriefs in Neuroscience,
DOI 10.1007/978-1-4614-5559-2_1, © Springer Science+Business Media, LLC 2013

of which 3 nM is independent of peripheral sources with the serum contribution being at most 14 [9]. The basal level of hippocampal estrogen was pegged at 8 nM, which remained fairly steady independent of castration. Their data thus indicate that estradiol in the hippocampus does not originate from peripheral testosterone. In females, estradiol levels in this structure are far lower than males (0.5–1.7 nM, lows reached at diestrus 1, highs at proestrus), but still higher than the 0.008–0.12 nM found in the plasma. Since peripheral sources of estradiol account for so little of the portion of native levels of the steroid, most or all of estradiol's direct actions in the hippocampus arguably stem from neural sources. An additional consideration is that the levels of steroids may be even more concentrated at synapses where they are produced and act.

Neurosteroid levels also differ regionally. For instance, allopregnanolone levels are 75% higher in the olfactory bulb than the hippocampus of adrenalectomized (ADX) and gonadectomized (GDX) rats [4]. Their synthesis is evolutionarily conserved across vertebrate nervous systems (reviewed elsewhere [10]). Region-specific changes in neurosteroids allow for selective, precise, and rapid changes in neuronal function. Neurosteroids are thus poised to regulate CNS functions with a cell-type specificity not attainable by peripheral steroids.

Neurosteroids have a broad spectrum of actions that overlap those ascribed to neuroactive steroids [3, 11]. Over the last 15 years, our understanding of their actions has exploded. This review highlights research on the role of neurosteroids and their synthesis in the mammalian pituitary and nervous system, particularly for the rodent and human. Peripheral steroids are referenced to the extent that they instigate changes in neurosteroids or have overlapping functions as yet undistinguished from neurosteroids. The neuroactive steroids mentioned are primarily gonadal, since most research in the area has focused on this source. This review also discusses clinical ramifications.

2 Steroidogenic Enzymes in the Nervous System

All steroids are made from cholesterol (Fig. 1). The source of cholesterol in the brain is de novo synthesis. The enzymes that generate steroid localize to specific cell types in the nervous system (reviewed in [3, 10]) and include the two proteins that initiate steroidogenesis – cytochrome P450scc and the steroidogenic acute regulatory (StAR) protein.

2.1 P450scc, StAR, and Neurosteroidogenesis

The presence of P450scc marks sites capable of synthesizing steroid. The enzyme catalyzes the formation of the parent hormonal steroid to all others, pregnenolone, from cholesterol in the inner mitochondrial membrane [12]. Substrate cholesterol is supplied to P450scc from the outer mitochondrial membrane through the action of StAR [13]. Hence, co-expression with StAR marks sites of ongoing steroid synthesis

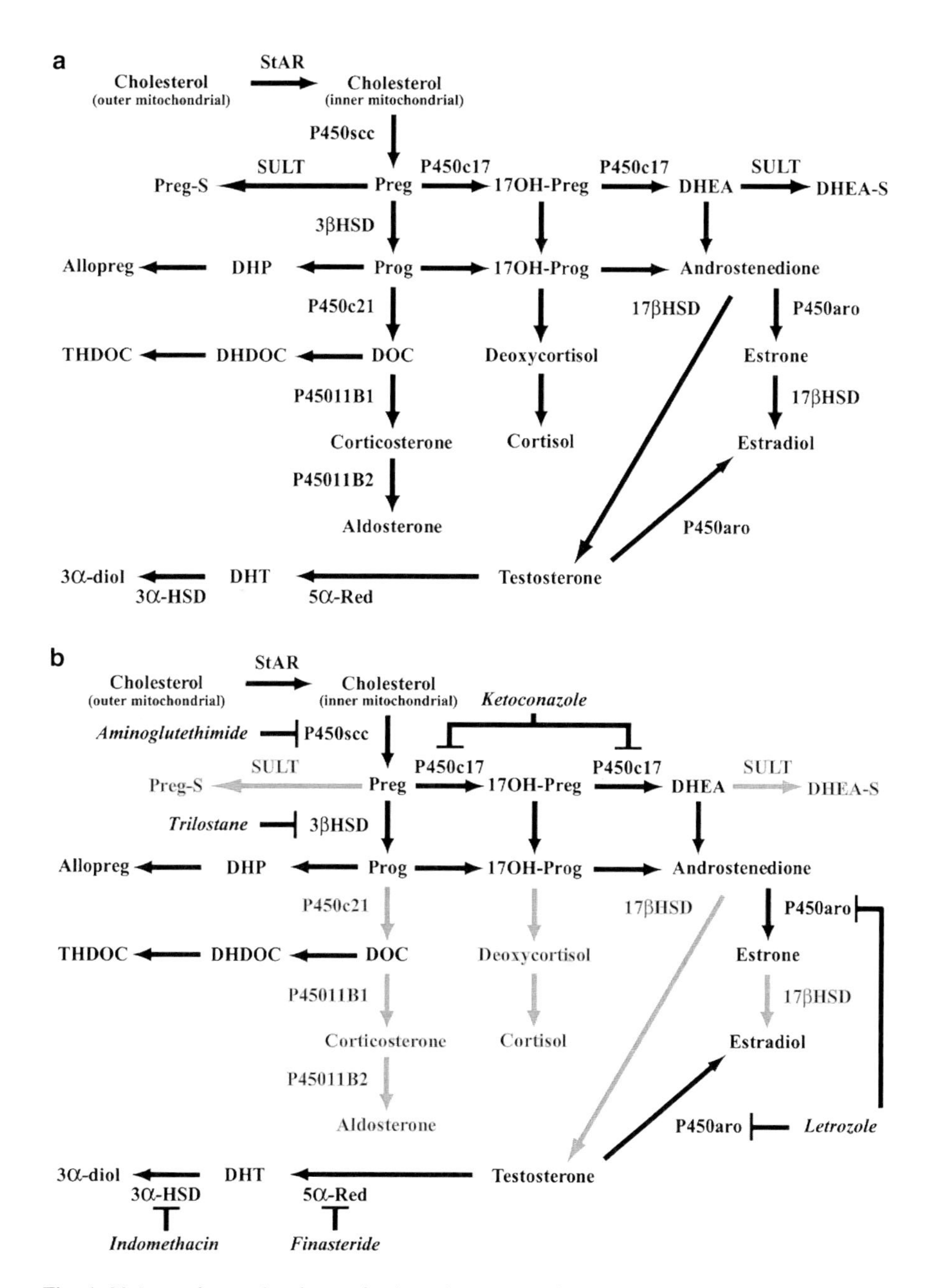

Fig. 1 *Major pathways for the production of neurosteroids and commonly used inhibitors.* (**a**) Steroidogenic pathways starting with the common rate-limiting step – the delivery of cholesterol from the outer mitochondrial to the inner mitochondrial membrane and P450scc by StAR. The specific family members of steroidogenic enzymes aside from P45011B1 and 2 are not labeled. Not shown is 3α-hydroxy-4-pregnen-20-one (3αHP), the 3αHSD metabolite of progesterone. (**b**) The actions of select enzyme inhibitors on steroidogenic pathways. Note that some inhibitors can be promiscuous; indomethacin opposes cyclooxygenase and aminoglutethimide can affect other P450 enzymes. Also of importance is that lipophilic inhibitors do not necessarily require direct infusion to enter the brain, as they can cross the blood–brain barrier [842–844]. Abbreviations are *OH* hydroxy, *5αRed* 5α-reductase, *SULT* sulfotransferase, *P450aro* aromatase, *HSD* hydroxysteroid dehydrogenase, *Prog* progesterone, *Preg* pregnenolone, *3α-diol* androstanediol. P45011B2 is also known as aldosterone synthase

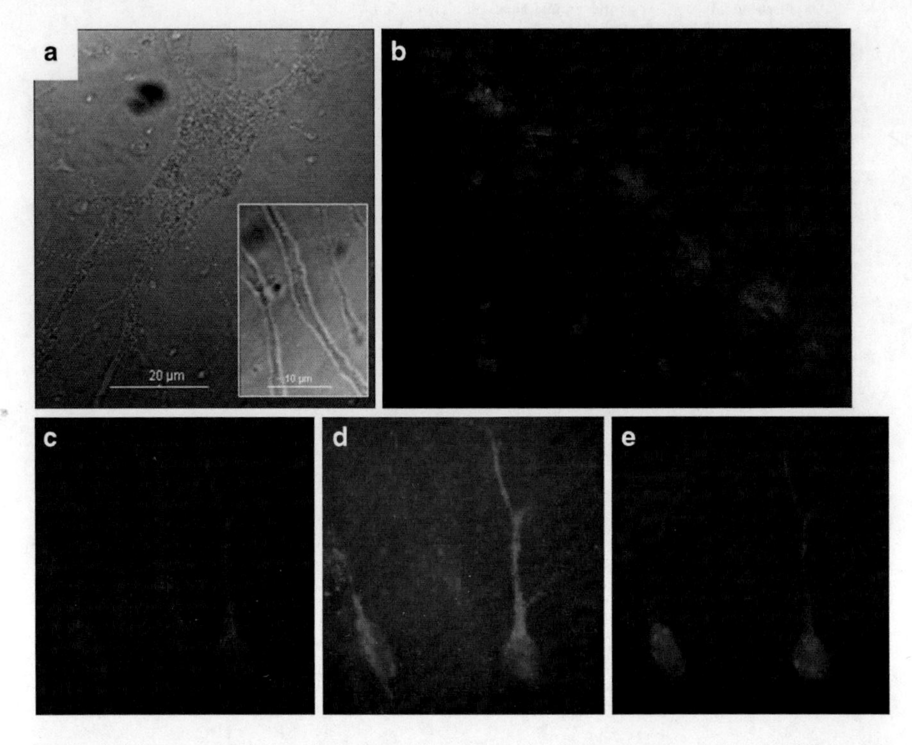

Fig. 2 *StAR expression in steroidogenic cells in the brain.* (**a**) Immunoreactivity of StAR (*red*) in a single hippocampal neuron in culture obtained from a P5 rat. *Inset*, Higher magnification image of StAR immunoreactivity in dendrites. (**b**) Large Purkinje cells in an adult murine cerebellar section are brightly painted (*orange brown*) with immunofluorescence for both StAR (*red*) and $P450_{scc}$ (*green*) with nuclear exclusion (DAPI, blue). Some co-expression is also present in neighboring cells. (**c–e**) Immunofluorescence for StAR (**c**, *red*) and $P450_{scc}$ (**d**, *green*) reveals mitochondrial-type colocalization (*orange brown*) with nuclear exclusion in two-layer V adult mouse cortical neurons; visualized in all three channels with blue DAPI stain in **e**. (**b–e**) 400× magnification (Figure 2a courtesy with permission from [461] and **b–e** adapted with permission from studies undertaken in [21])

(Fig. 2). Elimination of P450scc or StAR results in a catastrophic loss of steroids ([14, 15]; reviewed in [12]).

Steroidogenesis in the brain as a whole occurs at a much lower level than in endocrine tissues. Correspondingly, levels of StAR in the rat brain are approximately 0.1–1% that of the adrenal and P450scc, at least an order of magnitude lower [16–18]. A later survey in the human put the number for P450scc mRNA between 1,000 and 100,000 times lower, depending on the region of the CNS, with highest abundance in the spinal cord [19]. Rat spinal levels of StAR mRNA are approximately 5,000- and 37,000-fold lower than in the testis and adrenal gland, respectively [20].

Both proteins colocalize to individual neurons and glia within the developing and adult brain [16, 17, 21, 22]. Steroidogenic regions of the brain they identify include the cortex, hippocampus, hypothalamus, and the cerebellum, especially in Purkinje cells [21, 23–25]. Neural cells that produce steroid include neurons, astrocytes, spinal tract motor neurons, sensory neurons and satellite glia in dorsal root ganglia (DRG), and, as mentioned later, immature oligodendrocytes [3, 17, 21, 23, 26–28]. In the human fetal spinal cord and DRG, P450scc expression is confined to cell bodies [29]. Immature Schwann cells also produce P450scc, StAR, and steroid [30–32]. Contrary to earlier indications, neurosteroidogenesis is more widespread in neuronal populations than glia [21]. Both StAR and P450scc as well as steroid primarily localize to neurons [21, 24, 33], StAR mRNA being higher in gray matter than white [34].

Dual-label immunohistochemistry finds P450scc in rat microglial cells within the CA3 hippocampus using an injury model [35]. A separate study finds no evidence of the enzyme in resting or activated primary microglia from neonatal mice, but did observe a few other steroidogenic enzymes including StAR [36].

Embryonically, P450scc is detected in similar areas of the brain as in the adult and includes the rodent developing retina, superior cervical ganglia, DRG, trigeminal ganglia, and facio-acoustic preganglia as well as in embryonic day 9.5 (E9.5) to 11.5 neural crest cells, Rathke's pouch, and the neural tube [23]. Discrete cell populations in the adult rat retina synthesize StAR, P450scc, and hence steroids [37, 38]. A preliminary report uncovered expression of P450scc along with aromatase in select type II sensory cells in the circumvallate papillae of the rat taste bud [39]. StAR expression but not P450scc as of yet is also reported in pituitary gonadotropes [40].

2.2 *Peripheral Benzodiazepine Receptor/TSPO*

Some studies promote a role for the peripheral-type mitochondrial benzodiazepine receptor/translocator protein (PBR/TSPO) in steroidogenesis by itself or in combination with StAR [41]. This ubiquitous protein usually localizes to the mitochondria as a component of the permeability transition pore complex that controls mitochondrial integrity [42]. An involvement of PBR in steroid production is controversial. Expression of PBR is unregulated by steroidogenic stimuli, and its knockout is embryonic lethal, indicative of an essential function outside of steroidogenesis [43]. Still, endozepines reportedly bind mitochondrial PBR to regulate de novo neurosteroid synthesis in isolated mitochondria from rat C6-2B glioma cells [44]. However, PBR ligands also induce mitochondrial swelling and promote apoptosis, which affects steroidogenesis too [42, 43, 45]. Knockdown of PBR further impairs protein import, perhaps by compromising the mitochondrial electrochemical potential – another factor crucial for steroid synthesis [46–49]. In toto, evidence for this regulator of mitochondrial integrity having a direct rather than supportive or permissive role in steroid production remains elusive.

2.3 StAR-Independent Steroidogenesis

Steroid synthesis does occur in the absence of StAR. This unregulated minor pathway is primarily important in progesterone production by the human placenta, the only steroidogenic tissue known to not express StAR [12]. In cells that do express StAR, StAR-independent steroidogenesis contributes to basal steroid levels. The mechanism by which this occurs is unknown, but one pathway involves oxysterols. Hydroxylated forms of cholesterol freely cross mitochondrial membranes to P450scc without the help of StAR. In the brain, the presence of 24S-hydroxycholesterol represents a type of oxysterol that can conceivably be converted to steroid [50, 51]. Such conversion could have physiologic effects. In the testis, oxysterol-dependent steroid production is developmentally important [52].

2.4 DHEA, Sulfotransferase (SULT), and Sulfated Steroids

Steroid sulfatase activity also exists in the CNS, making sulfated steroids available for further metabolism [3]. In this way, pregnenolone sulfate can increase central allopregnanolone levels [4, 53, 54]. However, conflicting data exist as to the presence of sulfated steroids and steroid SULT activity [1–3, 6, 55–57] (Fig. 1a). The classic steroid SULT, SULT2A1, is in the rat brain [58]. The enzyme counts pregnenolone, DHEA, allopregnanolone, and glucocorticoids among its substrates. Recently, low levels of a brain- and testis-specific enzyme, SULT2B1a, were also identified in the rat brain and in C6 glioma cells [59, 60]. This α-amino-3-hydroxy-5-methyl-4-isoxazole-propionic acid (AMPA) receptor-inhibited enzyme selectively acts on pregnenolone, generating pregnenolone sulfate in glioma cells. The presence of either enzyme has yet to be confirmed in the human. Another problem is that the level of sulfotransferase activity in the brain may be too low to be physiologically relevant [61]. Hence, some suggest that sulfated steroids like pregnenolone sulfate exclusively originate in the CNS from the circulation [62]. However, a transport mechanism for charged steroids to cross the blood–brain barrier remains ill defined, but may happen [63].

A more troubling question raised by Liere and colleagues surrounds the authenticity of DHEA-S and pregnenolone sulfate detected in the brain [6, 63]. They conclude that prior assays inflated the levels of DHEA, pregnenolone, and sulfated derivatives due to issues with cholesterol autoxidation during assay purification steps [6, 7]. When that is taken into account, sulfated steroids and even DHEA are below detectable limits in the male rat brain. Still, they detect steroid sulfates in the human brain due to local synthesis or blood vessel contamination [63, 64]. Other groups maintain that sulfated steroids exist in the rodent brain, and certainly, the former studies do not absolutely dismiss the possibility [55, 65]. If sulfated steroids are absent, then what can be made of studies that show variations in central sulfated steroids and the role of SULT2B1a and SULT2A1? Could changes in oxysterols be what these assays truly measure? Moreover, is the finding that sulfated steroids impact CNS function serendipitous?

Given these uncertainties, the reports of sulfated steroids in the nervous system in this review are taken at face value. However, the reader should keep in mind the caveat that in many cases, the identities of the measured compounds are in question.

3 Regulation of De Novo Neurosteroidogenesis

Our knowledge of the signaling molecules and intracellular mechanisms that govern neurosteroid production is incomplete, but they commonly involve activation of cAMP and calcium pathways along with increases in and activating phosphorylation of StAR ([21, 66–73]; reviewed in [10, 74]) (Fig. 3). In a notable exception, robust cAMP stimulation causes a model of immature sciatic nerve Schwann cells to partially differentiate, accompanied by a decline in StAR mRNA levels [27, 30, 75]. Low levels of StAR mRNA decline in isolated microglia with prior exposure of the parent mouse to lipopolysaccharide (LPS) or increase upon co-incubation with the bacterial compounds and interferon-γ in vitro [36]. Oddly, dbcAMP has no effect on the expression of any steroidogenic enzyme including StAR in these cells either.

Unlike StAR, P450scc levels change more slowly with stimulation making them a less accurate indicator of acute variations in steroid production. In one study,

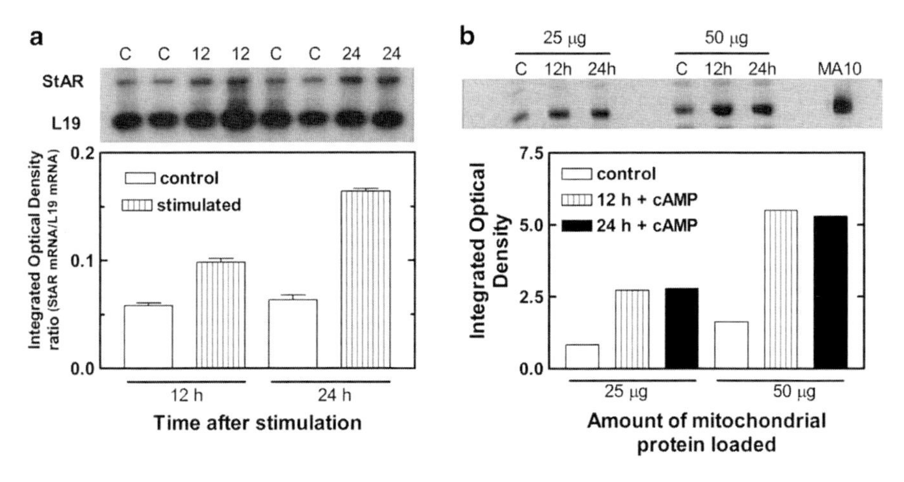

Fig. 3 *cAMP stimulates glial StAR mRNA and protein expression.* (**a**) StAR mRNA levels as detected by semiquantitative RT-PCR analysis (*top panel*, quantified in *lower panel*) increase 170% and 260% from basal compared to L19 ribosomal protein mRNA with 40-μM forskolin stimulation for 12 and 24 h, respectively ($p < 0.0001$ by one-way ANOVA; $p < 0.01$ by the Newman-Keuls multiple comparison test versus control). (**b**) Likewise, 0.5-mg/ml dibutyryl-cAMP (dbcAMP) stimulates StAR protein levels by 330%, remaining elevated over 24 h by immunoblot analysis (*top panel*, quantified in *lower panel*). *Last lane*, positive control for StAR, mitochondrial protein from stimulated mouse MA-10 Leydig tumor cells (Images courtesy with permission from [21])

dbcAMP and forskolin failed to increase P450scc in primary cerebellar granule cells after 12 h, but did elevate StAR, aromatase, and, by implication, estradiol synthesis [69].

Few compounds are known to instigate neurosteroidogenesis in mammalian systems. Recent in vitro and in vivo studies implicate a resident lipid in the CNS, palmitoylethanolamide (PEA). PEA upregulates StAR and P450scc through the peroxisome-proliferator-activated receptor (PPAR)-α in the rat brain, rat C6 glioma cells, and murine astrocytes to stimulate allopregnanolone production [76, 77]. Chronic stimulation with 1-μM 9-*cis* retinoic acid (RA) acting through the retinoid X receptor (RXR) similarly increases StAR, P450scc, and progesterone production in a human glial cell line [78].

As in the gonads, luteinizing hormone (LH) increases StAR expression in human M17 neuroblastoma cells and differentiated rat primary hippocampal neurons within 30 min [79]. In human SH-SY5Y neuroblastoma cells, gonadotropin-releasing hormone (GnRH) indirectly trebles StAR and P450scc levels inside 90 min by upregulating LH [80]. While receptors for both hormones are expressed in the human and rodent CNS, whether these hormones regulate StAR in vivo in the brain is unclear [79, 81–83]. As mentioned later, estrogen also increases steroid production through a mechanism unique to neural cells.

Neurotransmitters further regulate neurosteroid production. For instance, γ-aminobutyric acid (GABA) acting through $GABA_A$ receptors can inhibit pregnenolone, progesterone, and THDOC production [84–86]. Steroid production increases in the rat retina and in hippocampal sections with *N*-methyl-D-aspartic acid (NMDA) [87, 88] and in the rat retina with benzodiazepines [89]. Endozepines may act as an inverse agonist on the $GABA_A$ receptor and through plasma membrane-localized PBR to stimulate steroid synthesis [10, 90].

4 Steroidogenic Pathways

Enzymes downstream of P450scc shape the type of steroid produced by a cell. Pregnenolone generated by this enzyme is conjugated to sulfate or metabolized to progesterone by 3βHSD or 17OH-pregnenolone by P450c17 [3, 91, 92]. Further metabolism yields steroids like allopregnanolone, estradiol, and DHEA.

Most or all steroidogenic enzymes are represented in the CNS, generally at levels 2–5 orders of magnitude lower than adrenal or gonadal tissues [3]. Since not all cells express P450scc, an enzyme's presence does not always indicate participation in de novo steroid synthesis. Rather, the enzyme may simply convert a steroid originating from the circulation or a neighboring cell to another, such as gonadal testosterone to estrogen by brain aromatase or glial progesterone to allopregnanolone by neurons [93].

Moreover, one cell type can generate several different steroid end products. Select hippocampal neurons retain a full complement of enzymes to generate de novo progesterone, allopregnanolone, and estradiol, among other steroids [10, 93, 94].

Fluctuations in downstream enzymes impact the final steroid produced (this subject is not discussed in depth here, but is examined in [10]). Treatment for 48 h with 1-μM 9-cis-RA does not change StAR and P450scc mRNA in rat postnatal day 10 (P10)–P12 hippocampal slices [95]. However, P450c17, testosterone, and estradiol levels do escalate, suggesting a switch in the steroid end product. In general, the capability and capacity of a neural cell to produce a particular steroid changes with development, the ovarian cycle, or pathologic conditions.

5 Mechanism of Action of Neurosteroids

How neurosteroids affect target cells depends upon the receptors present. Steroids exert long-term genomic effects through classic nuclear steroid receptors, like androgen (AR), mineralocorticoid (MR), and glucocorticoid receptors (GR) and estrogen receptor α (ERα) and ERβ. Progesterone recognizes A and B isoforms of the nuclear progesterone receptor (PR) in the brain [96]. These steroids all bind their receptors at low to less than nanomolar concentrations. Androstanediol, allopregnanolone, and THDOC however do not activate PRs, but the latter two will be following reversion to DHP and DHDOC [97]. Allopregnanolone and pregnanolone also recognize the nuclear pregnane X receptor/steroid and xenobiotic receptor (PXR/SXR) at micromolar levels [98].

Gonadal and adrenal steroids are generally considered to exert mostly long-term genomic effects since changes in their production in the periphery do not reach target cells as swiftly as neurosteroids. Hence, research on neurosteroids has focused on their ability to elicit rapid, nongenomic changes that occur within seconds to milliseconds. This requires local synthesis of the steroids to generate the necessary concentrations in a prompt manner. The rapidity of the effect of neurosteroids is due to their stimulation of G protein-coupled and ligand-gated ion channel membrane receptors [99].

That said, not all important changes induced by neurosteroids are rapid. For instance, increases in the decay time constant of spontaneous inhibitory postsynaptic currents (sIPSCs) in hypothalamic magnocellular neurons by allopregnanolone require incubation with the steroid for at least 7–11 min [100]. Persistent stimulation by neurosteroids elicits both posttranslational and genomic effects through membrane and nuclear receptors that, for instance, change cell responsiveness to steroids. Conversely, peripheral steroids also use nongenomic pathways to elicit changes in the nervous system.

The number of membrane receptors available to neurosteroids is extensive. The following briefly describes these receptors with the caveat that direct binding by steroids has not been firmly established for many. A second note is that local concentrations of neurosteroids at synapses where they may be produced, for example, are expected to be much higher than regional levels. Since these local concentrations are not known, what constitutes a physiologic versus a pharmacologic dose presently can only be guessed.

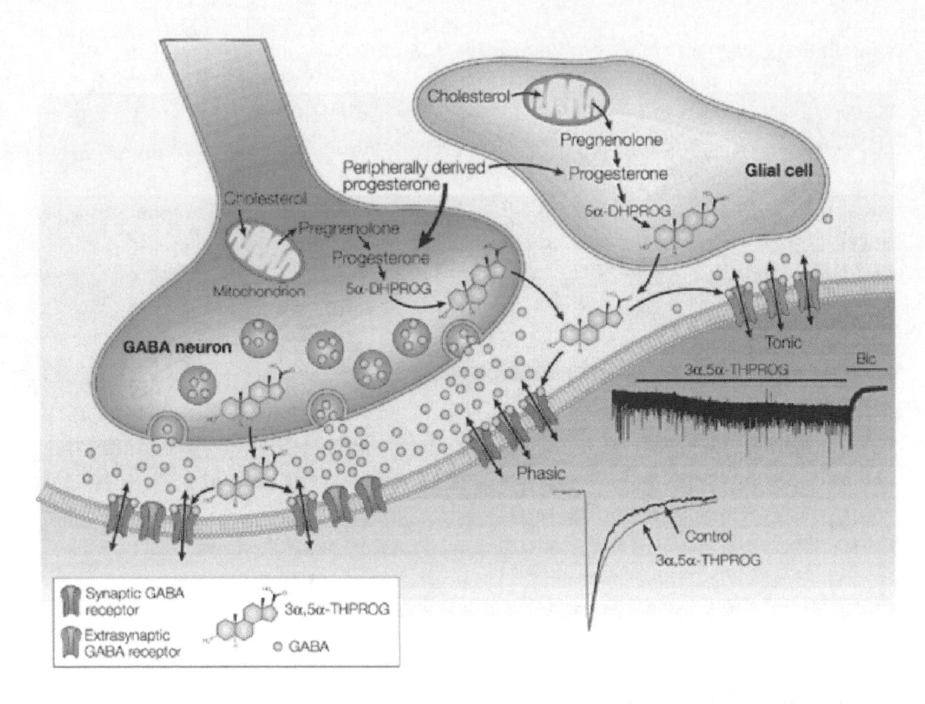

Fig. 4 *Modulation of $GABA_A$ receptors by allopregnanolone (THPROG) at an inhibitory synapse.* GABA release from synaptic vesicles activates postsynaptic $GABA_A$ receptors, generating a transient miniature IPSC (mIPSC) (phasic response). Steroids produced by local neurons or glia prolong the decay of this response and thereby enhance synaptic inhibition. Neurons can also contain extrasynaptic receptors, activated by ambient GABA to elicit a tonic background current. Under voltage clamp, this current is characterized by a noisy baseline that is uncovered by $GABA_A$ receptor antagonist bicuculline (Bic). In some neurons, it is enhanced by low levels of neurosteroids that otherwise have little effect on synaptic receptors thanks to the added presence of the δ subunit. DHPROG; DHP (Adapted and reprinted with permission from [101])

5.1 $GABA_A$ and $GABA_\rho$ Receptors

Select steroids bind the $GABA_A$ receptor, directly activating or allosterically modulating GABAergic neurotransmission [101–104]. They affect phasic and tonic inhibition through synaptic and extrasynaptic receptors, respectively (Fig. 4). Nanomolar levels of 5α-reduced steroids like androstanediol, allopregnanolone, and THDOC potentiate $GABA_A$ channel chloride currents, prolonging channel open time and reducing neuronal excitability with a 20- and 200-fold higher efficacy than benzodiazepines and barbiturates, respectively [102, 105–107].

The sites through which steroids accomplish this are now better understood. The $GABA_A$ receptor is a pentamer consisting of two α and two β subunits and a single γ subunit [108]. GABA itself binds in a specific pocket between an α and a β; thus, each receptor presents two potential binding sites. Allopregnanolone and THDOC potentiate receptor activity through a conserved site in the α subunit [109, 110]

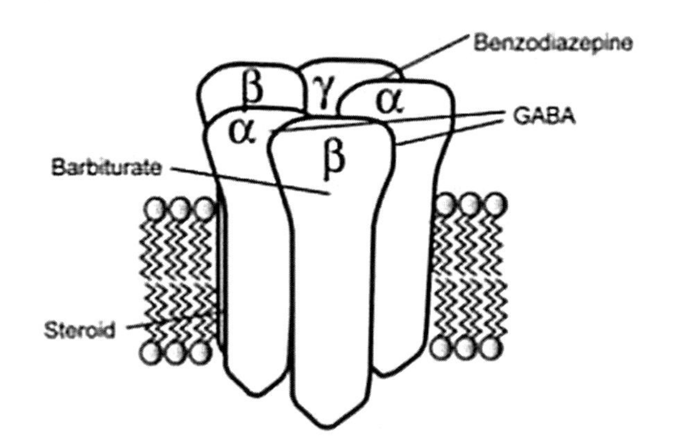

Fig. 5 *Points of regulation of the $GABA_A$ receptor.* The figure roughly depicts only one of the two steroid potentiating sites in the α subunit. Not shown are additional activating and inhibitory sites utilized by steroids (Adapted and reprinted with permission from [845])

(Fig. 5). Interaction with a single α is sufficient to potentiate GABA binding to either α or β site with consequent channel activation [111, 112]. At higher concentrations, these steroids directly activate the receptor through a separate interfacial site formed by the α and β subunits [109]. Binding to the potentiation site greatly enhances the ability of neurosteroids to activate the receptor. Pregnanolone has similar effects to allopregnanolone [113], but may utilize a different receptor interaction [114].

DHEA tends to weakly inhibit this receptor [115], while pregnenolone sulfate and DHEA-S noncompetitively and allosterically antagonize $GABA_A$ channels [116–118]. Pregnenolone sulfate inhibits channel activity through a third site [119–122].

Steroid sensitivity changes with age or specific conditions, like pregnancy [101, 103]. Differences in $GABA_A$ receptor responsiveness are dictated by factors such as subunit composition and phosphorylation [101, 123]. The α and γ subunits are particularly relevant, whereas the ε subunit promotes neurosteroid insensitivity [124–126]. Receptors containing α5 are less sensitive to allopregnanolone and THDOC [124, 127]. On the other hand, the general view is that allopregnanolone potentiates $GABA_A$ receptors containing the α_1 and γ subunits [128–131]. Meanwhile, an amino acid difference between the α4 and α1 subunits permits the steroid to rapidly desensitize α4β2δ receptors [132]. While the γ subunit is not required for neurosteroid responsiveness, γ1 confers less sensitivity than γ2 [101, 103, 124].

Substitution of γ with the δ subunit in extrasynaptic receptors increases tonic inhibitory conductance induced by allopregnanolone and sensitivity to other neurosteroids as well like THDOC, while losing responsiveness to benzodiazepines [103, 124, 133, 134]. The α4 subunit tends to co-express with δ [135], and the presence of both in α4β1δ recombinant receptors reduces the EC_{50} for GABA over α1β2γ2 [136]. Loss of the δ subunit alters $GABA_A$ receptor function and composition, increasing γ2 and decreasing α4 in regions of the brain that formerly express δ [135]. As with many other $GABA_A$ receptor subunits [137, 138], neurosteroids regulate the expression of the δ subunit and alter the localization profile of receptor populations [139]. Allopregnanolone and estradiol i.p. both increase δ levels after 48 h in rat CA1 hippocampus. The regulation of subunit composition is mediated in part through $GABA_A$ receptor stimulation [137]. In vitro, 100-nM allopregnanolone increases cell surface content of α4β2δ by 30 min in a GABA-dependent manner [140].

Given the potential therapeutic applications, a synthetic class of steroids called epalons has been developed. These analogs of allopregnanolone, like the natural steroid itself, modulate $GABA_A$ receptors without exerting hormonal genomic effects through classic nuclear receptors. An example is the anesthetic ganaxolone, a well-tolerated and more stable 3β-methylated form of allopregnanolone.

Steroids also influence related $GABA_\rho$ ($GABA_C$) receptors that contain the ρ1 subunit but not ρ3 [101, 141]. 5α-reduced allopregnanolone and THDOC allosterically potentiate this receptor through a specific domain in ρ1 [142]. This same domain is used by inhibitory 5β-reduced steroids, like pregnanolone. Notably, allopregnanolone works at micromolar levels instead of the nanomolar range needed for $GABA_A$. Estradiol also inhibits ρ1-containing receptors [143].

5.2 Glycine Receptors

Glycine-induced currents are enhanced by allopregnanolone concentrations as low as 1 μM [144, 145]. Progesterone, estradiol, and pregnanolone on the other hand oppose glycine-gated chlorine channel activation [106, 144, 146–150]. Pregnenolone sulfate and DHEA-S are even more potent inhibitors. The mechanism by which this occurs differs between steroids. Progesterone is a partial noncompetitive antagonist, while pregnenolone sulfate and, more sensitively at low to submicromolar levels, pregnanolone competitively inhibit glycine receptor activation [144, 145]. Estradiol inhibits the receptor at a separate site from pregnanolone at micromolar levels [150]. Both steroids additively reduce glycine-evoked currents.

The actions of individual neurosteroids are again determined by subunit composition. Allopregnanolone and pregnenolone enhance α1 glycine receptor currents [106, 147]. Pregnenolone has no effect on α2. Progesterone inhibits recombinant α2 receptors, but has no effect on α1 and α1β receptors expressed in *Xenopus* oocytes [147] or potentiates glycine induction of α1-containing receptors in expressed in heterologous human embryonic kidney HEK293 cells [151]. Estradiol inhibits receptors containing α2 and α2β, while the presence of α1 lends insensitivity to the steroid [150]. DHEA-S inhibits α1 and α2 glycine receptors at lower concentrations than α4 [147]. Thus, switches in subunit composition that occur with development alter the effects neurosteroids have on glycine receptors.

Steroid efficacy also changes with the presence of the β subunit. Estradiol has no effect on a homomeric α3 receptor, but inclusion of the β subunit renders the receptor susceptible to inhibition by the steroid [150]. The β subunit reduces the potencies of pregnenolone sulfate and DHEA-S as well as potentiation by pregnenolone [147].

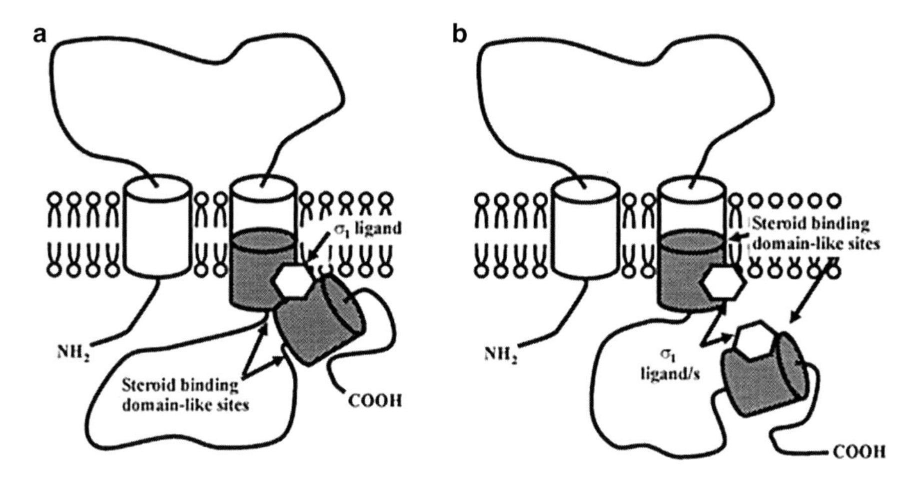

Fig. 6 *Model for neurosteroid binding of σ1 receptors*. Open cylinders represent the two putative transmembrane domains. Closed cylinders represent the steroid-binding domain-like sites, and the open hexagon represents a putative σ1 ligand. (**a**) Possible spatial arrangement of the ligand binding site involving both SBDL sites. (**b**) Alternative model for ligand interaction (Adapted and reprinted with permission from [846], copyright (2008) Bentham Science Publishers)

5.3 σ Receptors

The σ receptor family was originally identified based on function not homology, as a subtype of opioid receptor with antidepressant properties [152, 153]. These intracellular receptors transduce signals via physical interactions with other proteins, such as ion channels [152, 154]. The best-characterized receptor subtypes are the σ1 and σ2/PR membrane component-1 (PGRMC1) receptors.

5.3.1 σ1 Receptors

Ligand activation causes σ1 receptor migration from its usual endoplasmic reticular localization to other intracellular compartments [152]. Candidate endogenous ligands for this receptor include steroids, sphingolipids, and N,N-dimethyltryptamine (DMT) [152]. Testosterone, pregnenolone, pregnenolone sulfate, DHEA, DHEA-S, and progesterone competitively bind the metabotropic receptor [155–158]. The binding sites deduced from the receptor's homology with steroid-binding proteins like 17βHSD type 2 are formed by two steroid-binding domain-like (SBDL) regions [158, 159] (Fig. 6).

Pregnenolone sulfate and DHEA-S stimulate presynaptic σ1 receptor activity, potentiating the frequency of miniature excitatory postsynaptic currents (mEPSCs) [156, 160]. The receptor is also potentiated by DHEA [161].

Steroids utilize σ1 to modulate NMDA receptors. DHEA induces potentiation of NMDA-induced excitation of rat CA3 pyramidal neurons through σ1 receptors

[161, 162], and in hippocampal sections, 30-nM DHEA-S similarly potentiates NMDA-stimulated norepinephrine release [163]. Progesterone and testosterone competitively suppress potentiation by DHEA-S, but do not alter NMDA activity themselves [162, 163]. Micromolar levels of pregnenolone sulfate may increase postsynaptic NMDA and AMPA receptor activity in hippocampal slices through sensitization of presynaptic α7 nicotinic acetylcholine receptors (nAchR) or σ1 receptor activation, both of which can increase the probability of glutamate release [160, 164, 165].

While pregnenolone sulfate can stimulate the σ1 receptor, some studies indicate that it may act rather as an inverse agonist or have little or no activity. Submicromolar levels (100 nM) of the steroid inhibit σ1 receptor facilitation of NMDA-induced norepinephrine release in hippocampal sections, while progesterone blocks this inhibition [163]. In the CA3 hippocampus, i.v. pregnenolone sulfate and pregnenolone do not alter σ1 receptor activation [162].

Steroids may also influence receptor levels and/or receptor occupancy. ADX and GDX mice exhibit 49% higher receptor binding in the hippocampus and 77% in the cortex, suggesting a negative influence by peripheral steroids [166]. Finasteride (Proscar/Propecia) reverses these increases to control intact levels, indicative of a role for 5α-reduced neurosteroids and/or a negative influence by progesterone derived within the CNS (Fig. 1b).

5.3.2 PGRMC1 Receptors

PGRMC1 is a progesterone target variously localized to the endoplasmic reticulum and nucleus [154]. This membrane-associated receptor also known as tetrachlorodibenzo-dioxin-induced 25-kDa protein (25-Dx)/ratp28/hpr6.6 is present in the brain and spinal cord [167, 168]. PGRMC1 can complex with serpine 1 mRNA-binding protein 1 (SERBP1; plasminogen activator inhibitor 1 RNA-binding protein, PAIRBP1), binds heme, and was recently found to be synonymous with the pro-apoptotic σ2 receptor [154, 169]. Progesterone binds to and antagonizes ligand activation of PGRMC1 and σ1 receptors with equivalent nanomolar affinity and with consequent sodium current inhibition in HEK293 cells [170]. Other more functionally oriented studies suggest that progesterone activates PGRMC1 [171, 172]. Testosterone also exhibits affinity for the receptor [173]. However, steroid binding may not actually be direct as the authors of one study suggest [172], but instead be mediated through an as-yet unidentified protein (reviewed in [174]). The true natural ligand for this receptor remains unknown.

A related receptor, PGRMC2, is also expressed in the brain [171, 175, 176].

5.4 Ionotropic Glutamate Receptors

Steroids allosterically modulate ionotropic glutamate receptors. Pregnenolone sulfate and pregnanolone sulfate bind and inhibit AMPA receptor activity through the glutamate binding site of the glutamate receptor 2 (GluR2) subunit [148, 177]. Pregnenolone sulfate noncompetitively reduces kainate and AMPA-induced currents [148, 178, 179], while progesterone potentiates kainate receptors [180].

NMDA receptors contain distinct potentiating and inhibitory binding sites for steroid, which have nanomolar to micromolar effective concentrations. Steroid sulfates are the most potent. Micromolar pregnenolone sulfate potentiates the NMDA receptor, augmenting rises in intracellular calcium induced by NMDA, while pregnanolone sulfate generally allosterically inhibits opening of the cation channel only following receptor activation [102, 148, 178, 181–183]. Pregnenolone sulfate is more potent when applied prior to glutamate, increasing the amplitude of glutamate response up to fivefold and decreasing deactivation twofold [184, 185]. These effects rely upon NMDA receptor subunit (NR; glutamate NMDA receptor, GluN) composition. Potentiation of NR1/NR2B involves a steroid modulatory domain in the NR2B subunit [186, 187], a site distinct from that of pregnanolone sulfate [183, 188]. Pregnenolone sulfate potentiates NMDA, glutamate, or glycine stimulation of receptors containing NR1/NR2A and NR1/NR2B and inhibits NR1/NR2C and NR1/NR2D combinations, likely at the same site as pregnanolone sulfate [185, 188]. Pregnanolone sulfate inhibits the latter two subunit combinations with twofold more potency than the former [183]. While these are the prevailing effects of pregnenolone sulfate, it can also have the opposite effects in vitro, albeit weakly [185]. An exception is found in CA1 pyramidal neurons from postnatal rats younger than P6. Here, the steroid potentiates NR2D-containing receptors, possibly due to developmental differences in subunit composition [65]. Thus, pregnenolone sulfate modulates NMDA receptor activity directly at micromolar levels like pregnanolone sulfate or indirectly at nanomolar levels through, for instance, possibly the σ receptor [163].

DHEA-S at 100 μM slightly potentiates the receptor [178, 179]. Johansson and coworkers describe the existence of a third site at which nanomolar levels of pregnenolone sulfate promote and pregnanolone sulfate reduce NMDA receptor inhibition by the selective NR1/NR2B inhibitor, ifenprodil [189, 190]. Similar allosteric modulation by DHEA-S and allopregnanolone sulfate of the effects of ifenprodil is also reported [191].

DHEA also activates the NMDA receptor [192]. Glycine additively increases the response to DHEA in rodent embryonic neocortical neurons [192], perhaps reflecting DHEA modulation of NMDA activity through σ1 [161]. Estradiol is slightly potentiating at 20–100 μM on NMDA receptors [178, 179]. Estrogen also enhances NMDA receptor transmission in hippocampal neurons and directly inhibits NMDA receptor currents [193–195]. Estradiol mainly through ERs elevates or reduces select NMDA and AMPA receptor subunits in a region-specific manner [196–199]. While allopregnanolone naturally opposes NMDA through $GABA_A$ receptor modulation, there is evidence it supports NMDA receptor activity in select tissues, probably also indirectly [200, 201].

5.5 Transient Receptor Potential (TRP) Superfamily

Neurosteroids interact with TRP members. DHEA competitively and to a lesser degree, pregnenolone sulfate and DHEA-S inhibit vanilloid receptor 1 (VR1)/

TRPV1/capsaicin receptor activation in DRG while estradiol strongly potentiates it [202, 203]. Pregnenolone sulfate acts noncompetitively. Pregnenolone sulfate, pregnanolone sulfate, pregnanolone, DHT, and especially progesterone inhibit TRP canonical 5 (TRPC5) [204]. DHEA-S and estradiol are weaker inhibitors.

TRP melastatin 3 (TRPM3) cation channels bind in order of strength, pregnenolone sulfate, progesterone, and DHT [205, 206]. Micromolar levels of progesterone inhibit and pregnenolone sulfate strongly activate TRPM3 to regulate glutamate release at Purkinje cells from neonatal rats [205, 207]. Progesterone, DHT, and to some extent pregnanolone and estradiol reduce activation [205]. DHT competitively inhibits pregnenolone sulfate activity at high concentrations (>1 μM), while progesterone at levels as low as 10 nM directly inhibits TRPM3 activation either through the pregnenolone sulfate-binding site or a separate site altogether.

5.6 Membrane Estrogen Receptors

Over the past decade, numerous new and familiar steroid receptors have been identified that localize to cell membranes. Estrogen has fast nongenomic effects on glia and neurons through membrane-localized ERα and ERβ. One mechanism of action for ERα is through interacting with metabotropic glutamate receptors (mGluR) at the plasma membrane. Current data suggest that this interaction only occurs in females [208]. A separate intracellular membrane-associated protein, G protein-coupled receptor 30 (GPR30; G protein-coupled estrogen receptor 1, GPER1), preferentially binds 17β-estradiol and stimulates extracellular signal-regulated kinase (ERK; mitogen-activated protein kinase, MAPK) activity through the epidermal growth factor receptor (EGFR) [209–212]. This stands in contrast to ERα which binds both 17β and 17α isomers of estradiol and inhibits ERK activity [213].

The existence of yet another ER, ER-X, has been proposed. ER-X localizes to cellular membranes of murine P2 neocortical neurons and preferentially responds to 17α-estradiol, which, as an aside, binds ERα poorly [213]. Stimulation of ER-X promotes protein kinase C (PKC) activation of ERK1 and ERK2 pathways at picomolar levels.

Estradiol rapidly inhibits G protein-coupled metabotropic $GABA_B$, dopamine D2, and μ and κ opioid receptors [214–218]. The effect on D2 receptors may be indirect via ER-mediated changes in receptor number or G protein-coupled inwardly rectifying K^+ (GIRK) channel expression [219, 220].

Estradiol may directly interact with opioid and $GABA_B$ receptors in vivo [221]. These inhibitory effects may rely on uncoupling of μ opioid and $GABA_B$ receptors from GIRK channels by activation of GPR30 [209–211]. A study on arcuate neurons from GPR30-knockout mice suggests that these effects involve a second uncharacterized $G\alpha_q$-coupled membrane ER (mER-$G\alpha_q$) that stimulates phospholipase C (PLC), PKC, and protein kinase A (PKA) activities [222]. This second receptor is characterized by the effects of a synthetic ligand that is an analog of tamoxifen, STX. The variety of estrogen receptors therefore permits estrogens to have complex effects on neuronal functions.

5.7 Other Membrane Steroid Receptors

Progesterone activates a Gi protein-coupled membrane PR family (mPRα, β, and γ), also referred to as progestin and adipoQ receptors (PAQR 7, 8, and 5) [223–225]. Isoform B of the nuclear PR can localize to the plasma membrane and directly activate c-Src and hence ERK-1/2 [226]. While PR-A binds Src as well, its predominant nuclear localization suggests a limited role in rapid intracellular signaling. AR can localize outside the nucleus too and interact with Src at the plasma membrane, leading to ERK activation [227, 228]. Whether this has effects aside from those on gene expression remains to be explored.

A long-postulated membrane AR (mAR) that stimulates PLC activity and elicits transient increases in intracellular calcium levels [229, 230] was recently identified as the promiscuous G protein-coupled receptor 6A (GPRC6A) [231]. Unlike AR, GPCR6A is far more sensitive to testosterone than DHT in a heterologous cell line (~20 vs. 200–400 nM). The receptor apparently is also sensitive to 60-nM estradiol. Progesterone at 40 nM exerts an inhibitory effect. Loss of the receptor in vivo eliminates the nongenomic actions of testosterone [231].

5.8 Other Receptors

Pregnenolone and progesterone noncompetitively inhibit muscarinic receptors [232–235], while allopregnanolone, progesterone, testosterone, and pregnenolone sulfate allosterically inhibit neuronal nAchR [236–240]. One study found that pregnenolone sulfate potentiates presynaptic α7nAchR activity and, through it, L-type voltage-gated calcium channels (VGCCs) in hippocampal slices [164]. Progesterone, corticosterone, and estradiol also inhibit ganglionic $\alpha3\beta4$ receptors [241]. Estradiol binds and positively allosterically modulates human neuronal nAchR ($\alpha4\beta2$) while inhibiting the receptor in the rat [242].

Various steroids allosterically influence a family of ionotropic ATP receptors, purinergic P2X receptor channels (reviewed in [243]). Estradiol, estrone, and DHEA-S inhibit P2X receptors in rat pheochromocytoma PC12 cells. Inhibitory effects are also described for testosterone, corticosterone, and cortisol. At the same time, DHEA selectively potentiates receptors containing the $P2X_2$ subunit in rat DRG neurons [244]. Progesterone has similar effects, but only on $P2X_2$ homotrimers [245]. Select steroids in the hundreds of nanomolar range allosterically modulate $P2X_4$ receptors, the most abundant subtype in the brain. Pretreatment with allopregnanolone or THDOC augments ATP-evoked currents by recombinant receptors in transfected cells, while pregnanolone reduces these currents [246]. THDOC additionally increases deactivation time. Unfortunately, binding studies in support of a direct interaction by neurosteroids on P2X receptors are currently lacking.

Micromolar estradiol, testosterone, allopregnanolone, and progesterone allosterically and noncompetitively antagonize serotonin/5-hydroxytryptamine (5-HT) type 3 receptors [247]. Estradiol also directly binds and activates the sodium-activated Slack (sequence like a calcium-activated K^+) channel [248] and β1, 2, or 4 subunit-containing large conductance calcium-dependent potassium channels (BK; maxi-K; Slo1; KCa1.1) [249, 250] and binds and potentiates L-type VGCCs in vitro [251]. Notably, estradiol diminishes activated L-type VGCC currents. Estradiol acts indirectly through membrane ERα/mGluRs in the absence of glutamate to oppose channel activity as shown in rat hippocampal pyramidal neurons, but as noted before, solely in females [208].

Testosterone, DHEA, and glucocorticoids also activate BK channels [250]. A recent study indicates that estradiol closes open BK channels [252]. Pregnenolone sulfate is implicated in directly activating VGCCs [253] and α_{1B}-adrenergic receptors in the prelimbic cortex [254] and potentiating murine inwardly rectifying potassium Kir2.3 channels [255]. 5α-reduced steroids like allopregnanolone block T-type calcium channels [256].

Allopregnanolone is thought to influence dopamine D1 receptors in the rat midbrain ventral tegmental area (VTA), probably indirectly through $GABA_A$ [257, 258]. At the same time, D1 antagonists block ~15-μM DHEA-S increases in EPSCs and glutamate release in rat prelimbic cortex slices, while a σ1 receptor antagonist achieves only a partial block [259]. A follow-up study however finds that 1-μM DHEA-S inhibits 5-HT-induced glutamate release in prelimbic cortex fractions through the σ1 receptor, not through any direct effect on D1 [260]. Progesterone may similarly utilize this receptor to contest D1 activity [261]. Thus, direct functionally relevant binding of the D1 receptor by steroids seems unlikely.

Progesterone competitively inhibits G protein-coupled oxytocin receptors [262–265] at nanomolar to micromolar levels [266]. This effect partly depends on the presence of the G protein. Most studies suggest progesterone directly binds the receptor, and in humans unlike the rodent, it may not be the most efficacious progestin. Notably, not all studies agree that progestins actually bind and inhibit this receptor [266].

DHEA also binds neurotrophin nerve growth factor tyrosine kinase receptor TrkA and $p75^{NTR}$ receptor in PC12 cells [267], while 30–50-nM pregnenolone binds microtubule-associated protein 2 (MAP2) [268]. An embryonic isoform MAP2c also binds pregnenolone and DHEA [269, 270]. Molecular modeling reveals a homologous domain in the protein for binding DHEA similar to that of 17βHSD type 1 [270].

6 Neurosteroid Functions

Through this wide range of receptor targets, neurosteroids impact many different functions in the CNS and PNS. These compounds are implicated in development, neuroprotection, and a variety of behaviors. Too numerous to mention are the concomitant widespread changes brought about by peripheral steroids. What follows are highlighted areas of basic and clinical research that implicate neurosteroids or where distinctions between peripheral and local steroid actions are blurred.

7 Neuronal Growth, Synaptic Plasticity, and Brain Development

Maternal, placental, and fetal steroids and postnatal peripheral steroids dramatically affect brain development and function. Unfortunately, few studies have effectively parsed roles for neurosteroids from those entering the CNS from the periphery in utero. The proposition that effects on neural precursor proliferation and maturation in adult tissues are primarily mediated by neurosteroids is more easily justified.

7.1 Neurosteroid Production

The ability of the brain to synthesize neurosteroids is acquired early on [22, 23, 91]. Embryogenesis is the time of greatest neuronal growth and development and is accompanied by the expression of P450scc in many different structures as mentioned previously [23]. Levels of neurosteroidogenic LH also rise in differentiating embryonic rat primary cortical neurons [83]. The presence of StAR has yet to be rigorously examined in the embryo.

In the postnatal CNS, StAR is produced in dividing cells and germinal layers [24]. In the male rat hippocampus, P450scc mRNA peaks between P4 and P14, well after maternal and placental steroids are essentially cleared from the bloodstream, before dropping severely, while levels of StAR remain constant to adulthood [271]. Developmental peaks in StAR and P450scc mRNA are reached in the rat cerebellum around P10 [69]. These peaks are higher in males than in females.

Developmentally, neurons, glia, and oligodendrocytes possess differing abilities to generate steroid. Astrocytes express all of the enzymes necessary to synthesize pregnenolone, DHEA, and sex steroids, as shown for those in the neonatal rat cerebral cortex [272]. Neurons obtained from this region of the CNS also generate pregnenolone, progesterone, DHEA, androstenedione, and estrone. Neonatal oligodendrocytes however only express P450scc and 3βHSD. Human embryonic neuronal progenitor Ntera2/cl.D1 (NT2) cells secrete pregnenolone [273]. With RA-induced differentiation to a fetal neuronal-like cell type (NT2-N), cells now generate progesterone with the onset of 3βHSD expression. Both NT2 and NT2-N cells can further metabolize progesterone to DHP and, to a lesser extent, pregnanolone and allopregnanolone.

Early reports show that cultured oligodendrocytes from perinatal rats express P450scc and produce more and more steroid over time [274–276]. As well, P450scc mRNA is detectable in cDNA arrays from the oligodendroglial cell line N19, a model for immature oligodendrocytes [277]. Other studies fail to show the presence of P450scc or StAR in this cell type in vivo or from mixed glial populations [17, 23, 24]. It now appears that rat oligodendrocyte pre-progenitors generate progesterone, but sequentially lose this ability with maturation [28]. Production of 7α- and 20α-hydroxylated forms of pregnenolone may remain intact, albeit at a lower level.

7.2 *Functional Effects of Neurosteroids*

7.2.1 Progesterone, Testosterone, and Estradiol

Neonatal peaks in StAR, P450scc, and aromatase mRNA in Purkinje cells and other cerebellar neurons reflect high levels of de novo synthesis of progesterone and estradiol that regulate dendritic growth, spine generation, and synaptogenesis in an autocrine/paracrine manner [69, 278–280]. Progesterone's actions are mediated through antiprogestin-sensitive PRs [281] (Fig. 7). Allopregnanolone has no influence on these events. Progesterone also increases the proliferation of neural progenitor cells from the adult rat dentate gyrus independent of 5α-reductase activity [171]. This effect of progesterone is mediated by the PGRMC1 receptor.

Estradiol i.c.v. promotes dendrite growth and synapse and spine formation in neonatal Purkinje cells and corrects the deficits in such parameters in aromatase-knockout mice [278]. Estradiol rapidly reverses the regression of hippocampal neurite extensions [282] and increases neurite length in E14 rat cortical neurons in an ERα-independent manner [283]. This steroid has many other effects on neuronal structure and plasticity as well [284]. Synaptic plasticity and developmental effects are at least partly due to rapid and longer term genomic actions through ERα and β, for instance, as compared in the hippocampus of ERα- and β-knockout mice [285]. A further role in neuronal development is reflected by the observation that expression of the putative estrogen receptor ER-X peaks in P7–P10 mouse immature neocortical neurons [213]. Although the steroid is protective for neurons, in cultured P3 rat astrocytes, >40 h 100-nM estradiol inhibits ERK1/2 activation and increases cell death especially in females via ERα [286]. It also inhibits proliferation selectively for female-derived cells due to gender-specific expression of the ERα. On the other hand, 10-nM estradiol as a 24–48-h pretreatment also through ERα protects against the death of P1–P3 mouse cerebrocortical astrocytes due to ischemia [287]. At the same time, testosterone may affect

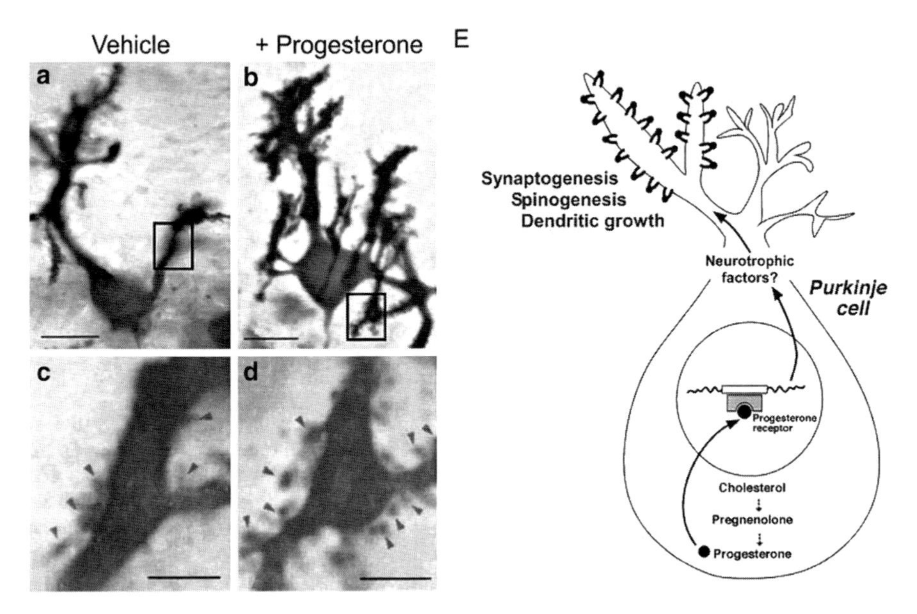

Fig. 7 *Changes in dendritic growth and spine formation in the developing Purkinje cell are elicited by progesterone.* Cerebellar cultures from neonatal rats were grown in the presence of ethanol vehicle (**a**) or 100-nM progesterone (**b**) and immunostained for calbindin. Dendrites treated with the steroid show improved development and at the ultrastructural level as observed in the *boxed regions* in **b** and **a**, greater densities of spines (*arrowheads*) (**d**) over control (**c**), respectively. Potential mechanisms for organizing actions of progesterone produced in the developing Purkinje cell. *Scale bars*: **a**–**b**, 20 μm; **c**–**d**, 5 μm. (**e**) Model for progesterone promotion of Purkinje cell maturation in the neonate through an autocrine mechanism. Newly synthesized neuroprogesterone stimulates PR-mediated transactivation of genes for neurotrophic factors that promote developmental changes (Adapted with permission from [280, 847])

the developing cerebral cortex through increases in $GABA_A$ receptor α2 subunit levels [138].

7.2.2 DHEA and DHEA-S

Recent data propose a role for endogenous DHEA production in neuronal proliferation and differentiation in the murine developing neural tube [288]. This hypothesis is bolstered by the fact that P450c17 expression initiates in the developing nervous system by E9, prior to its appearance in the placenta and testis which otherwise could supply exogenous DHEA (the mouse fetal adrenal does not generate DHEA). Whether neural DHEA arises from actual de novo production, conversion from circulating precursors, or maternal supply remains unclear.

Both DHEA and DHEA-S at levels as low as 10 nM promote neuronal development and survival as well as glial survival or differentiation in cultures from E14 mouse embryos [289]. DHEA increases axon length and the formation of varicosities and basketlike processes of neurites from the embryonic neocortex possibly through NMDA receptors [192]. DHEA-S on the other hand increases dendrite length independent of NMDA receptors [192]. DHEA also increases the proliferation of human neural stem cells isolated from the fetal cortex through σ1 and consequent activation of NMDA receptors [290] and s.c., growth, and survival of nascent neurons in the rat dentate gyrus while opposing diminution of their number by corticosterone [291].

7.2.3 Pregnenolone and Pregnenolone Sulfate

Ultrastructural changes in neurons can involve direct binding of microtubules by neurosteroids with subsequent effects on microtubule organization. In this manner, 500-nM pregnenolone induces microtubule assembly in vitro [268] and, at 40 μM, protects against its disruption in PC12 cells by nocodazole [292]. The former effect is opposed by progesterone at the same level (and pregnenolone sulfate), though 30 μM of either steroid increases neurite outgrowth in vitro [268, 292].

Pregnenolone sulfate stimulates neuronal proliferation in the dentate gyrus and enhances expression of polysialylated neuronal cell adhesion molecules (NCAMs) in young and aged mice [293]. Neurosteroids like pregnenolone sulfate may also developmentally modulate neural cell plasticity, growth, and maturation using AMPA receptors. For pregnenolone sulfate, actions through presynaptic NMDA receptors result in glutamate release and, ultimately, enhance postsynaptic AMPA receptor activity in CA1 pyramidal neurons from rats younger than P6 [65]. It also stimulates glutamate release at neonatal Purkinje cells through TRPM3 channels [207]. In concert with a role in differentiation, the steroid decreases proliferation of E18 rat hippocampal neurons in culture as well as reportedly increasing the length and number of neurites [294].

7.2.4 Allopregnanolone

The sensitivity of $GABA_A$ receptors to neurosteroids fluctuates during development with subunit expression. As noted earlier, neurosteroids can be the architects of such changes. Allopregnanolone, for instance, regulates changes in select α and β subunit mRNA levels in neonatal cortical neurons in vitro, pointing to a developmental role for the steroid [137]. Sensitivity to the steroid increases in rat dentate granule cells during development, concomitant with modifications in subunit composition [295].

Allopregnanolone induces alterations in neuronal populations and localization in the prefrontal cortex and thalamus in neonates, resulting in changes in behaviors that depend upon the prefrontal cortex [296–298]. Allopregnanolone increases or reduces cell proliferation in the dentate gyrus at low (100–500 nM) or high (>0.1 μM) doses, respectively, and, in the former case, causes regression of fetal hippocampal neurite extensions without interrupting established connections with other cells [282, 293, 299]. This regression is part of a mitotic response in which coordinated changes in cell cycle genes lead to the proliferation of neural progenitor cells in E18 rat and human cerebral cortical neural stem cells and increase proliferation of murine HT-22 hippocampal cells [294]. While the minimal dose to enhance rat cell division is 100 nM, only 1-nM allopregnanolone is needed for human cells. Higher doses (100 μM) repress proliferation for the rat cells. 0.1- to 1-μM allopregnanolone also induces proliferation of immature rat cerebellar granule cells in vitro [300]. These genomic and proliferative changes are mediated through L-type VGCCs associated with $GABA_A$ receptors [294, 300].

In summary, while neuroactive steroids regulate brain development and differentiation, the coordinated regulation of neurosteroid synthesis may shape structural organization, neuronal proliferation, axonal growth, and synaptogenesis. Additional examples are detailed below. These functions have acute and long-term effects on brain functions and behaviors, such as anxiety.

8 Neurosteroids, Anxiety, and Psychiatric Disorders

8.1 Anxiety and Stress

Steroids regulate fear and anxiety and are implicated in anxiety disorders. Acute stress itself induces CNS and serum progesterone, allopregnanolone, and THDOC, with essentially all brain THDOC derived from adrenal sources [84, 301]. In a separate battery of tests of acute stress, anxiety coincides with a brief increase in cerebrocortical allopregnanolone and THDOC that occurred in some cases without changes in serum steroid [302]. Reduced anxiety with habituation results in no change in central allopregnanolone or precursor progesterone and pregnenolone in response to an acute stressor, and aged animals exhibited a higher spike in the steroid over younger rats.

The rise in central allopregnanolone is not wholly accounted for by peripheral sources as demonstrated, for example, in stressed 10-day post-ADX adult male rats [301]. This rise affects $GABA_A$ receptor activity and neuronal excitability in select regions of the brain.

8.1.1 3α-Reduced Steroids Can Promote Anxiolysis

Administration of allopregnanolone or THDOC induces anxiolytic, analgesic, and sedative effects [103]. For instance, intrahippocampal administration of 1.26-μM allopregnanolone in male rats reduces locomotor activity, taken as a measure of increased sedation [236]. Numerous studies document that the modulation of $GABA_A$

receptors by allopregnanolone is key for its anxiolytic and antidepressive effects [303]. The steroid also induces relevant genomic changes. Short-term allopregnanolone administration upregulates δ subunit expression, further sensitizing the $GABA_A$ receptor to the steroid's actions and increasing its anxiolytic effect [139].

Anxiolysis induced by allopregnanolone is at least partly due to $GABA_A$ receptors in central neurons in the amygdala [304]. Systemic or simply intra-amygdalar inhibition of 5α-reductase increases anxiety [305, 306]. Microgram amounts of the 5β-isomer of allopregnanolone, pregnanolone, injected into the dorsal hippocampus or lateral septum reduce measures of anxiety [307]. The $GABA_A$ receptor antagonist picrotoxin blocks these effects, reflecting the receptor's involvement.

Following up on studies that show a relationship between early life stress and allopregnanolone, rats bred for higher maternal separation anxiety exhibit lower hippocampal and amygdalar allopregnanolone and elevated anxiety (except proestrus females who exhibit no difference in the steroid, but do in anxiety) [308–310]. Administration of allopregnanolone counters the developed anxiety.

The effect of allopregnanolone on anxiety changes with gonadal steroid-dependent changes in $GABA_A$ receptor composition. Shifts in δ subunit levels in the dentate gyrus follow changes in the ovarian cycle in the rodent. Increased levels during late diestrus when serum progesterone is high are anxiolytic [311]. Supplemental allopregnanolone i.c.v. has no effect on anxiolysis in diestrus, but does increase open arm entries in the plus maze during the progesterone- and allopregnanolone-poor time of estrus [312]. Loss of the δ subunit reduces or eliminates the anxiolytic effect of allopregnanolone or ganaxolone [313, 314]. As discussed before, both estradiol and allopregnanolone upregulate the subunit as well. Moreover, early life loss of 3αHSD activity in the neonate via finasteride injections results in heightened anxiety as measured in the plus maze later in life [315]. The change in anxiety is not improved by hippocampal allopregnanolone supplementation, indicative of a developmental role for 3α-reduced steroids.

Chronic social isolation stress, a model of posttraumatic stress disorder, reduces 5α-reductase expression and allopregnanolone along corticolimbic circuits in rodents [316–318]. Inhibition of reductase activity or s.c. rescue with allopregnanolone doses that restore cortical, hippocampal, and amygdalar content of the steroid, respectively, increases or reduces the duration of contextual fear-conditioned freezing behavior in mice [318].

A recent outlier to this story comes from a study on animals given restraint stress for 30 min. While nanomolar to micromolar 3α-reduced steroids and pregnenolone normally oppose corticotropin-releasing hormone (CRH) neuronal activity [319, 320], one group recently found that systemic administration of 20 mg/kg i.p. THDOC prior to stress increases serum corticosterone through actions at the male mouse hypothalamus and that finasteride reduces measures of anxiety observed with immediate testing in the elevated plus maze over both stressed and unstressed animals [320]. This confusing result may reflect a different role for adrenal THDOC compared to neural allopregnanolone, which, as mentioned, even systemically opposes anxiety caused by restraint stress (e.g., at much lower levels, 4 mg/kg s.c. in the rat against shorter term stress, 5 min [321]).

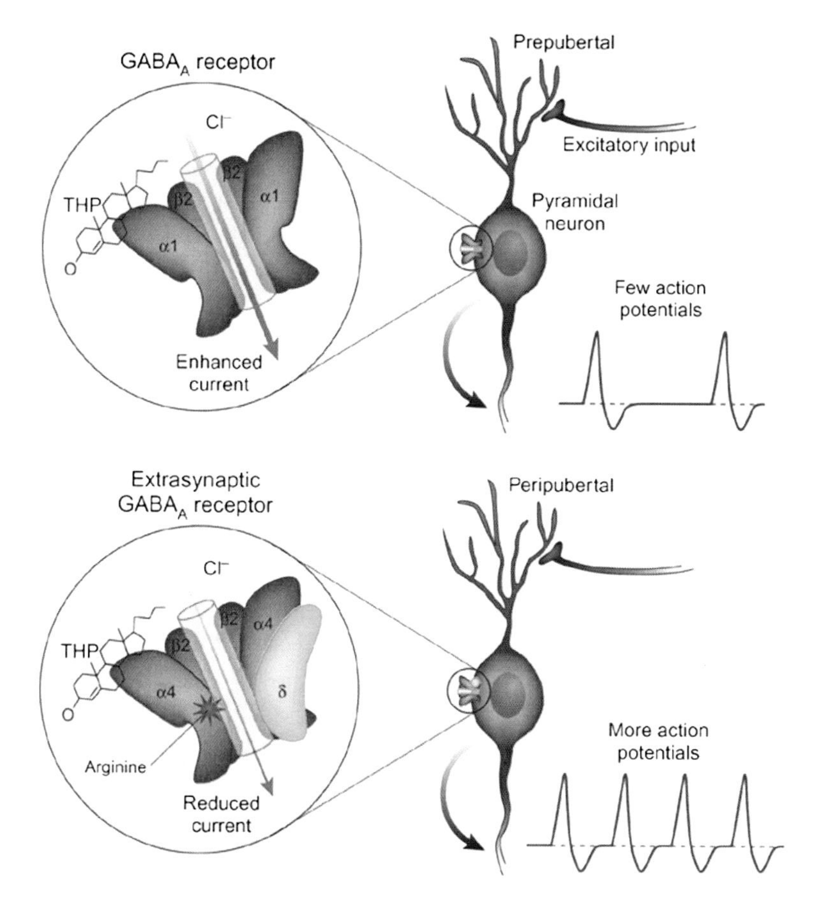

Fig. 8 *Model for paradoxical effects of allopregnanolone on anxiety.* Before and after puberty, allopregnanolone (*THP*) enhances GABA$_A$ receptor activity. During puberty in female mice, levels of extrasynaptic α4 and δ-containing GABA$_A$ receptors which mediate tonic inhibitory currents rise in the CA1 hippocampus. This change is linked to a drop in allopregnanolone. The receptor is now highly sensitized to the steroid due to the δ subunit. Due to the presence of an arginine residue in the α4 subunit, allopregnanolone inhibits these receptors, causing rapid desensitization and reduced outward inhibitory current. This excitatory effect on hippocampal pyramidal neurons leads to increased anxiety (Adapted and reprinted with permission from [848])

8.1.2 The Anxiogenic Effects of 3α-Reduced Steroids in Puberty and a Model of Anorexia

A landmark study by Shen and colleagues discovered that an increased proportion of α4- and δ-containing extrasynaptic GABA$_A$ receptors in CA1 pyramidal neurons in the pubertal hippocampus correlates with a switch in allopregnanolone from promoting anxiolysis to being anxiogenic in female mice [132] (Fig. 8). With the modification in receptor composition, the steroid changes from enhancing outward tonic currents to reducing them, raising input resistance and leading to increased neuronal excitability as tested in CA1 pyramidal cells. This paradoxical effect of allopregnanolone depends on the presence of both δ and α4, a subunit linked to increased anxiety [322]. No alterations in the allopregnanolone response at puberty are observed in δ subunit-null mice [132]. As well, mutation of a critical residue in α4 blocks outward current reduction by the steroid.

As mentioned before, $GABA_A$ subunit expression is regulated by steroids. The change in receptor composition correlates with a drop in hippocampal allopregnanolone with puberty [132]. If steroid levels are maintained rather than allowed to drop during puberty, the anxiogenic effect of allopregnanolone is lost. These findings provide a mechanistic insight into how allopregnanolone regulates the anxiety commonly experienced by teenagers.

A similar effect is found in an activity-based anorexia model. Affected adolescent female rats do not experience the postpubertal fall in α4 and δ subunits seen in controls with increased levels at CA1 spines [323]. This likely contributes to stress-based anxiety.

8.1.3 Estrogen, Progesterone, and Anxiety

Estradiol and, in some cases, progesterone are anxiolytic. Administration of estrogen or a selective ER modulator (SERM) with preference for ERβ decreases anxiety and depressive behavior in ovariectomized (OVX) mice, while an ERα SERM has no effect [324]. Similarly, diestrus mice exhibit higher anxiety compared to proestrus mice, but the same level as proestrus ERβ-nulls [325]. The relevant locations of ERβ activity include the bed nucleus of the stria terminalis and anterodorsal medial amygdala [326].

Interestingly, serum estrogen elevates serum and CNS levels of allopregnanolone [327–330], suggesting that estradiol and allopregnanolone cooperate to improve mood. Estradiol supplementation restores rat CNS and serum levels of allopregnanolone and decreases adrenal levels following ovariectomy, while co-treatment with the antiestrogen SERM raloxifene analog, which has affinity for both ER isoforms, only blocks the change in brain allopregnanolone [331, 332]. Thus, estradiol influences allopregnanolone synthesis from cholesterol or local or serum precursors in the CNS. Chronic administration of estradiol s.c. alternatively reduces serum and cerebrocortical and hippocampal pregnenolone, progesterone, and allopregnanolone in female rats [333]. This does not affect anxiety in the elevated plus maze [333].

On the other hand, social isolation stress selectively increases StAR, P450c17, 17βHSD, and aromatase mRNA and estradiol in the male rat hippocampus, but not components of the allopregnanolone synthetic pathway [316]. Thus, while peripheral estradiol can increase CNS allopregnanolone, locally produced estradiol apparently does not.

Generally, the anxiolytic effects of gonadal progesterone and estrogen are due to their regulation of serotonin 5-HT receptor activity, opioid receptors, and β-endorphin release [327]. For progesterone, amelioration of anxiety in rat models of immobilization stress involves the metabolism of the steroid to allopregnanolone and subsequent effects on $GABA_A$ receptor activity [303, 334–336]. Similarly, anxiolysis induced by progesterone in a Vogel conflict test is eliminated by co-treatment with finasteride [306]. Loss of PR results in greater responsiveness to systemic progesterone as gauged in the elevated plus-maze test, while finasteride completely blocks progesterone-induced increases in open arm entries [337]. Thus, progesterone through PR opposes anxiety reduction by allopregnanolone, but promotes anxiolysis through its sequential reduction to the latter steroid.

An involvement of PRs in anxiolysis cannot be ruled out. Treatment with s.c. RU486 opposes progesterone-facilitated anxiolysis and increases anxiety in control rats [338]. Still, PR activation is more generally associated with anxiogenesis. For instance, progesterone and a synthetic PR agonist, medroxyprogesterone acetate (MPA), antagonize estradiol-supported reductions in anxiety in macaques [339]. Contraceptive use is associated with changes in mood, with the synthetic progestin component linked to lower serum and CNS pregnenolone, progesterone, and allopregnanolone and increased anxiety in female rats given the drug s.c. [333]. An associated decline in $\alpha 1$ subunit levels in the cerebral cortex could come with an increased proportion of $\alpha 4$-containing $GABA_A$ receptors that would account for the rise in anxiety. A preliminary study in fact links MPA to acute increases in $\alpha 4$ in the CA1 hippocampus [340].

Complicating the picture, separate research found that intrahippocampal finasteride inhibits allopregnanolone levels and anxiolytic activities while increasing exploration and social interactions by estrus rats [341]. The reason for increased exploration with the suppression of allopregnanolone is unclear, but may relate to the steroid's effects on sedation and locomotor activities and the effects of resultant increases in DHP and progesterone on PR.

8.1.4 The Opposing Actions of DHEA, Pregnenolone, and Their Sulfate Conjugates in Anxiety

Pregnenolone and its sulfate conjugate increase anxiety and oppose anxiolytic and sedative effects induced by benzodiazepines and alcohol [11, 315, 342, 343]. Interestingly, the effect of pregnenolone sulfate is dose dependent. I.p. administration of the steroid is anxiolytic in mice at low doses but anxiogenic at higher doses (0.1 vs. 1.0 μg/kg) in the plus maze [342]. Over 0.5 mg/kg pregnenolone sulfate i.p. is once more anxiolytic in a mirrored chamber test that measures approach-conflict behavior [303], and s.c. attenuates conditioned fear responses through a $\sigma 1$ receptor-sensitive mechanism [344]. Progesterone opposes the steroid's effects in the latter test. These data suggest different physiologic and pharmacologic effects for pregnenolone sulfate which may partly involve its metabolism to allopregnanolone.

DHEA reduces anxiety, for instance, as observed in the elevated plus-maze test [345]. Although the steroid itself is somewhat antagonistic to $GABA_A$ receptor activity, a receptor antagonist blocks its effectiveness in this test. DHEA also diminishes anxiogenic pregnenolone sulfate levels in GDX mice brains [346].

0.05–0.5 mg/kg DHEA-S i.p. increases open arm entries in the plus maze [345]. Consistent with a bimodal effect, 1 mg/kg has no effect on plus-maze performance and opposes the anxiolytic effect of ethanol [345]. 2 mg/kg DHEA-S i.p. is anxiogenic in the mirrored chamber test [303]. It further opposes anxiolysis by a noncompetitive NMDA receptor antagonist, dizocilpine. However, 50 mg/kg s.c. opposes the conditioned fear response in mice (i.e., blocks stress-induced declines in motor activity following a footshock session), while 25 is without a significant effect [344]. The positive effect of the former dose is blocked by a σ1 receptor antagonist and progesterone. The reason that an even higher dose of DHEA-S or, as mentioned above, pregnenolone sulfate emulates the anxiolytic effect seen at low doses albeit in a different test is unclear. Of possible relevance is that the study's authors note that post-shock animals display lower serum levels of DHEA-S than control animals. Another explanation would be that at high doses, a substantial amount of the steroid is converted to anxiolytic metabolites.

Developmentally, exposure of juvenile rats to stress produces long-lasting increases in anxiety in the adult along with increased DHEA-S in the hypothalamus and entorhinal cortex [347]. Unlike allopregnanolone, the effects of DHEA-S and pregnenolone sulfate in the rat do not involve the central nucleus of the amygdala [304]. Their actions involve σ1 receptors, with progesterone being antagonistic [344].

8.1.5 The Roles of Neurosteroids in Other Models of Stress

Stress conditions such as postoperative and swim stress rapidly increase DHEA-S in the brain and allopregnanolone in the cerebral cortex in the absence of changes in serum levels in ADX and ADX/GDX rodents [1, 301]. Pregnenolone and progesterone also increase in the brain and serum with CO_2 inhalation or acute footshock stress [348]. CO_2 stress causes immediate and transient increases in DOC and progesterone and longer term increases in allopregnanolone and pregnenolone in the cerebral cortex independent of changes in serum levels [349]. The rise in neurosteroids corresponds with a normalizing of anxiety and $GABA_A$ receptor activity post-stress.

Anxiogenic pro-convulsant drugs that negatively modulate $GABA_A$ receptors also increase levels of largely anxiolytic steroids pregnenolone, progesterone, and THDOC in the brain and the serum [84, 85, 302]. This presumably occurs to mitigate the drugs' effects or as part of the stress response since coadministration with an anxiolytic benzodiazepine-like drug i.p. precludes the increase in steroid [84, 85]. In certain cases, this change in CNS steroid levels may entirely originate from adrenal sources since it is abolished in ADX/GDX rats [302].

Neurosteroids are also prophylactic against stressors. Adaptation to footshock stress in the rat by i.c.v. allopregnanolone involves its inhibition of Ach release likely through $GABA_A$ receptors and is most effective when the steroid is administered 30 min prior to the stress [350].

8.2 Therapeutic Synthetic Drugs Increase Neurosteroidogenesis

Many antidepressants including morphine, selective serotonin reuptake inhibitors (SSRIs), and tricyclic antidepressants increase central allopregnanolone content [351]. This may be due to direct promotion of 3αHSD activity and a reduction in allopregnanolone metabolism [352, 353]. Sertraline (Zoloft) further inhibits the reverse oxidative reaction of 3αHSD [352]. A separate study disputes the idea of a direct SSRI effect on increasing forward enzymatic activity [354]. Whatever the mechanism is, SSRIs like fluoxetine (Prozac) and its metabolite norfluoxetine increase allopregnanolone in the brain and CSF [355]. Norfluoxetine specifically increases corticolimbic allopregnanolone and corrects changes in contextual fear-conditioned responses by socially isolated mice [318]. The SSRI paroxetine (Paxil) i.p. also substantially raises DHP after 9 day and allopregnanolone after 21 day in the murine cortex and hypothalamus, independent of serum changes as determined by thin-layer chromatography [356]. The olfactory bulb also sees elevated allopregnanolone content. Fluoxetine and, less potently, paroxetine i.p. rapidly increase hippocampal allopregnanolone within 30 min in ADX/GDX male rats [357].

The SSRI etifoxine also targets $GABA_A$ receptors and increases allopregnanolone in the CNS of ADX/GDX rats and the serum and CNS of intact male rats [306]. Etifoxine maintains its anxiolytic effect in ADX males [358]. I.c.v. allopregnanolone or i.p. progesterone increase the anxiolytic efficacy of etifoxine, while inhibition of 3βHSD, 3αHSD or 5α-reductase activity blocks the drug [306, 358].

Importantly, Pinna et al. note that SSRIs elicit anxiolysis at doses 50 times less than those required to affect 5-HT reuptake [355]. One caveat is that as mentioned before, allopregnanolone inhibits 5-HT3 receptor activity in a heterologous cell model [247]. However, it occurs at concentration (10 μM) orders of magnitude greater than the nanomolar effective dose of the steroid at $GABA_A$ receptors and that may not be realized by SSRI stimulation. Altogether then, these data strongly suggest that neural production of allopregnanolone underpin the anxiolytic effects of SSRIs. This conclusion carries over to many other antidepressants as well.

Drug support of CNS allopregnanolone production however may have a price. Abrupt discontinuation of antidepressants can produce withdrawal symptoms that include anxiety, panic, and aggression. Speculatively, a cut in allopregnanolone levels due to cessation of antidepressant use may alter $GABA_A$ composition as with puberty and contribute to the changes in mood. Maintenance of allopregnanolone levels through withdrawal therefore may alleviate the side effects.

Cerebrocortical allopregnanolone, THDOC, and pregnenolone can be induced by other anxiolytic drugs, like olanzapine and clozapine [85, 359–361]. This increase correlates with concomitant changes in serum progesterone and corticosterone in rodents and is not observed in ADX rats for both drugs and in ADX/GDX rats for clozapine, suggesting a serum origin for the increase in steroid [360, 362].

Benzodiazepines like clonazepam also increase pregnenolone production in isolated rat retinas through $GABA_A$ receptors [89], suggesting they may do so in other cell types. The steroid end product is not yet characterized, but it is unlikely to be allopregnanolone. Finasteride co-treatment fails to inhibit the anxiolytic effect of clonazepam in Vogel conflict test [306].

Mood improvement seen with the SERM DT56a (Femarelle) orally in menopausal women corresponds with increases in central and serum allopregnanolone in OVX rats similar to that obtained with oral estradiol [363]. Another SERM, raloxifene analog which acts as an antiestrogen only in the presence of estradiol, also increases the steroid in the brain but only in OVX rats [331].

Various progestins can also increase allopregnanolone levels. It is not clear that they do so through simple metabolism since regional changes elicited by such compounds are specific. The contraceptive nestorone only increases allopregnanolone in OVX rats in the presence of supplementary estradiol and, even then, only in the anterior pituitary in the absence of an increase in serum levels [329].

8.3 Depression

The function of neurosteroids in anxiety is part of a larger role in mood and adaptation to stress. Indeed, neurosteroid levels change with emotional state and treatment with steroids can be ameliorative. Pregnenolone and DHEA can improve mood and depression, with DHEA a palliative strategy for patients with mild depression [11]. Allopregnanolone may also be effective.

Levels of DHEA are reduced relative to controls in the prefrontal cortex, VTA, and nucleus accumbens in the Flinders sensitive line rat model of depression [364, 365]. Long-term pretreatment by i.p. or infusion into the VTA or nucleus accumbens of the steroid shortens immobility time in the Porsolt forced swim test, a measure of despairing behavior [364]. Its antidepressive effect is linked to $GABA_A$ receptor activity, with δ subunit levels higher in the VTA and reduced in the nucleus accumbens of the rat line compared to controls [364]. However, its effect is dose dependent. Infusion of 3-nM DHEA into either brain region is antidepressive, while 30 nM has no effect. Finasteride infusion into the amygdala of receptive female rats increases immobility time in the forced swim test, suggestive of a similar antidepressive role for 3α,5α-reduced steroids [305].

As indicated, elevated DHEA can be antidepressant for depressed and aging patients, including those with major depression [366–368]. Administration of pregnenolone sulfate or DHEA-S reduces immobility time in the forced swim [369, 370] and tail suspension tests [371]. In one study, pregnenolone sulfate only reduced immobility time in ADX/GDX males [370]. On the other hand, removal of peripheral steroid sources potentiates the effects of DHEA-S. The antidepressant effects of these sulfated steroids are blocked by σ1 receptor antagonists like progesterone or through increases in progesterone levels by inhibition of 5α-reductase, suggesting the involvement of metabolism of the steroid to allopregnanolone [369–371]. Potentially, the effect of DHEA-S further involves reversion to DHEA and subsequent effects by this latter steroid. Since serum progesterone rises during forced swim tests in intact but not ADX/GDX animals, it follows that the removal of peripheral sources of inhibitory progesterone potentiates DHEA-S and σ1 receptor agonist activity [370]. Indeed, the high levels of serum progesterone experienced during pregnancy are linked to reduced σ receptor-mediated potentiation of NMDA responsiveness in CA3 hippocampal neurons [372]. This suggests one mechanism by which pregnancy influences behavior.

Peripheral as well as neural steroids are linked to postpartum depression and anxiety. Studies in rats found a strain-dependent difference in whether estradiol or its withdrawal promotes depressive-like behavior in the forced swim test [373]. Sudden declines in estrogen levels such as postpartum in the rodent correspond with increased cortical DHEA-S and decreased sulfatase activity [374, 375]. Resulting changes in anxiety then partly reflect increases in sulfated neurosteroids that antagonize GABAergic activities. Another cause is related to changes in allopregnanolone as mentioned later.

It is not always clear whether the relevant source of steroids in depressive disorders is from the brain or the periphery. Reduced serum and CSF levels of allopregnanolone correlate with cases of depression, such as major depressive disorder [11, 376, 377]. A small clinical study found that fluoxetine or another SSRI, fluvoxamine, restores CSF allopregnanolone and pregnanolone to control levels by 8–10 week [377]. This time frame correlates with the alleviation of depression. The increase occurred in the absence of any change in pregnenolone or progesterone, whose levels were similar or a multiple higher than the combined 3α-reduced steroids. Allopregnanolone potentiates the effects of antidepressants and symptom improvement with antidepressants correlates with increased allopregnanolone [376, 378]. Infusion of the steroid in the CA3 hippocampus or the central amygdala alleviates depressive-like behavior in a stress-based learned helplessness test [379]. Its effects on the former structure are mediated by $GABA_A$ receptors, while the latter are paradoxically for allopregnanolone, linked to NMDA receptor activation [201, 379].

8.4 Premenstrual Syndrome (PMS) and Premenstrual Dysphoric Disorder (PMDD)

In humans, serum levels of allopregnanolone can correlate with PMS and a more severe form of PMS, PMDD [11, 380]. These disorders are marked in part by depression, anxiety, and irritability following menstruation and during the luteal phase of the ovarian cycle when serum progesterone is the highest. Administration of estradiol or progesterone to PMS patients with a leuprolide blockade of LH secretion leads to a recurrence of symptoms, whereas leuprolide alone or the same regimen given to unaffected women has no effect [381]. Neurosteroids may participate, since like gonadal steroids, their levels change with ovarian steroids as mentioned and the ovarian cycle [382], which in turn can affect anxiety.

During the luteal phase, $GABA_A$ receptor function is altered. This is difficult to model in rodents, since the comparable period is brief. Exposure to high levels of progesterone during the cycle lasts a few hours, whereas in women, this period is measured in days. To more closely mirror the situation in humans, intact female rats are given prolonged doses of steroid. Although 3α-reduced steroids are acutely anxiolytic, chronically high levels of pregnanolone (48 h) briefly increase anxiety [139]. This transient rise in anxiety is matched by a similar change in hippocampal CA1 α4 and δ. This may help explain the cause of early symptoms in PMS. At shorter time points (90 min), the development of allopregnanolone tolerance in an anesthesia model involves a reduction in α4 [383].

As noted, shifts in δ subunit levels follow the estrus cycle, with its abrupt changes in serum progesterone [311]. As progesterone levels fall in late diestrus, α4 and to some extent δ rise, for instance, in the periaqueductal gray [136, 384]. Given that α4 is further hostage to changes in allopregnanolone [322], PMS and related syndromes like postpartum depression begin to bear similarities to pubertal anxiety. Withdrawal of rodents from a longer term progestin supplementation regimen mimics the steep loss in progesterone seen with the end of the luteal phase or pregnancy. Such withdrawal in the female rat provokes an increase in α4, $GABA_A$ receptor insensitivity in CA1 pyramidal neurons to a benzodiazepine, and anxiety within 24 h [322, 385]. Chronic administration of progesterone with indomethacin (Fig. 1b) however blocks the manifestation of withdrawal effects, reflecting the central role of progesterone-supported allopregnanolone [385]. Estradiol facilitates the withdrawal effect of progesterone, possibly by further stimulating CNS allopregnanolone synthesis [139, 385]. These data point to an increase in δ and α4 in PMS.

Indeed, women with PMS display benzodiazepine insensitivity as assessed by saccadic eye movement velocity and self-rated sedation scores, with more severe forms of the syndrome being more refractory to treatment [386, 387]. This insensitivity suggests that δ subunit-containing $GABA_A$ receptor populations are elevated.

Could then a higher level of serum progestins set up changes in $GABA_A$ receptor composition that precipitate PMDD? When prepubertal female mice are given finasteride i.p. for 3 days and then challenged with allopregnanolone, they exhibit heightened anxiety as demonstrated in the elevated plus maze [314]. To some extent, this prolonged loss and sudden increase in allopregnanolone mirrors the changes that occur with the drop in progesterone with the start of the follicular phase and resumption of the steroid on entry into the luteal. The changes in anxiety are accompanied by increases in α4 and δ. Mice null for δ do not exhibit anxiogenesis upon allopregnanolone add-back in spite of displaying a lower but significant increase in α4. Withdrawal effects are also blocked when α4 expression is reduced by antisense [322]. Again, this points to a requirement for both subunits to generate the withdrawal effect and suggests that shifts in allopregnanolone causes similar changes in these subunits during the menstrual cycle. If levels of the steroid are abnormally high levels in PMS, these changes may be more robust.

Are the changes in CNS allopregnanolone due to the metabolism of serum progesterone? Low doses of progesterone that do not elicit a withdrawal effect in rats can do so if the animal is supplemented with estradiol from 1 day prior to 1 day after initiation of chronic progestin administration [385]. Similar facilitatory effects by estradiol are observed in a separate study by the same group [139]. Since estradiol can increase CNS allopregnanolone, it may make up for the shortfall in the latter steroid caused by the use of less peripheral progesterone. It raises the possibility that local production of allopregnanolone is relevant in PMS and PMDD.

The data also imply that the prevention of drastic changes in CNS allopregnanolone, and thus, $GABA_A$ subunit composition will oppose the manifestation of these syndromes. Symptom alleviation by SSRIs may involve such a mechanism through their stimulation of de novo central allopregnanolone synthesis. As would be predicted, continuous dosing regimens through the highs and lows of ovarian steroids during the cycle are the most effective in patients [388].

In total, the data indicate that sensitivity to allopregnanolone is key to the development of PMS and PMDD. Ovarian steroids are the primary determinant through their regulation of CNS allopregnanolone levels and hence disruptive changes in the $GABA_A$ receptor. The extent to which pathologic changes in allopregnanolone derive from peripheral or neural sources remains unclear, but stimulation of local synthesis through the luteal/follicular phase transition is likely palliative.

8.5 *HRT and Mood*

Steroid changes also occur with age and loss of gonadal function. Hormone replacement therapies (HRT) sometimes in combination with antidepressants may offer an option to improve mood and depression in patient populations such as postmenopausal women using estrogen [389] or aging men with testosterone [390]. The effects of SERMs and estrogen on allopregnanolone were previously mentioned as mechanisms for mood improvement. Indeed, paroxetine and, by inference, central allopregnanolone improve psychological symptoms in postmenopausal women [391]. Other studies find varying success with therapies that compensate for the age-related loss of serum DHEA and DHEA-S from the adrenal zona reticularis [11]. In this case, the steroid is thought to act directly on the CNS rather than through an increase in neurosteroid synthesis.

8.6 *Other Psychiatric Disorders*

Neurosteroids have potential roles in many psychiatric illnesses. Pregnenolone and DHEA levels are elevated in the parietal cortex and posterior cingulate of postmortem brains from patients with schizophrenia and bipolar disorder [8]. Reduced central expression of aromatase is linked to autism and related pathologies in the ataxic *staggerer* mouse [392]. Cerebellar estradiol moderates loss of Purkinje cells in perinatal heterozygous *reeler* male mice, which models several neurodevelopmental disorders like schizophrenia [393]. Neonatal administration of estrogen also diminishes some behavioral and neuropathologic symptoms [394].

Gender-specific differences in the appearance of autism, schizophrenia, and depression may partly derive from differences in the production of and responsiveness to neurosteroids [8, 395, 396]. For instance, cerebellar impairments are a feature of schizophrenia, a disorder that trends higher in males with earlier presentation. Interestingly, StAR, P450scc, and aromatase mRNAs reach developmental peaks in the cerebella of P10 male rats that well exceed those in females [69]. Another gender difference is observed with anxiety. A postnatal stress model reveals that adult males repeatedly exposed to i.c.v. allopregnanolone display less anxiolytic behavior (i.e., reduced grooming) than similarly treated females [308].

Neurosteroids like allopregnanolone, pregnenolone sulfate, DHEA, and DHEA-S hold promise for the treatment of schizophrenia. Antipsychotics like lithium, olanzapine, and clozapine alter central neurosteroid levels, which in turn modulate dopaminergic and GABAergic neurotransmission through NMDA, $GABA_A$, and σ1 receptors [11, 397]. Research links progesterone and allopregnanolone to changes in one model of schizophrenia, deterioration of prepulse inhibition (PPI). Infusion of 0.2-μg allopregnanolone but not pregnenolone sulfate into the dorsal CA1 increases PPI to an auditory cue [398]. Given the promising laboratory data and that some antipsychotics rely on increases in neurosteroids, preclinical studies with the common precursor pregnenolone were pursued by three groups [397]. Their collective findings show that the steroid improves important aspects of schizophrenia and schizoaffective disorder without significant side effects. Larger studies are now underway.

In a male rat model of obsessive-compulsive disorder (OCD), i.p. fluoxetine and i.c.v. allopregnanolone comparably reduce parameters of compulsive and persistent behaviors [399]. Correspondingly, restraint stress-induced attenuation of OCD behavior is sensitive to finasteride. Observations in aromatase-knockout males and cycling female rats further implicate estradiol, progesterone, and other steroids in the disorder [400, 401]. Interestingly as in PMS, reduced benzodiazepine sensitivity occurs in OCD patients and other conditions, such as panic disorder [402, 403]. Antidepressants that selectively increase CNS allopregnanolone have emerged as the treatment of choice in the clinic [404].

Peripheral steroids and gender together also play a large role in other mood disorders. Lower plasma levels of anxiolytic steroids correlate with disorders like panic attacks, aberrant childhood behaviors (attention-deficit hyperactivity disorder [ADHD], oppositional defiant disorder, conduct disorder), and eating disorders (hyperphagia, anorexia nervosa, and bulimia) [11].

Gonadal estrogens act centrally through ERs to indirectly mediate changes in food intake [405]. The involvement of neurosteroidogenesis in these changes is unclear. Hyperphagia is induced by i.p. pregnanolone and allopregnanolone in male rats, reflecting increased time eating new foods due to reduced neophobia and, thus, changes in appetitive aspects of feeding behavior [406, 407]. This response involves $GABA_A$ receptors and is opposed through receptor antagonism by pregnenolone sulfate [408]. This effect of allopregnanolone may be replicated by increases in serum levels of the steroid provoked by stressors and disorders, like anorexia.

In contrast to allopregnanolone, pregnenolone sulfate also evokes hypophagia. Generation of hypophagia by DHEA-S may involve modulation of NMDA receptors [408].

9 Neurosteroids and Cognition

9.1 *Promnesic Neurosteroids*

Like peripheral steroids, neurosteroids affect cognition. Pregnenolone and testosterone also display memory-enhancing effects in footshock active avoidance behavior tests, while progesterone and estradiol do not [409]. Other studies find that DHEA improves cognition in depressed and aging patients (e.g., [367, 368]). The effects of DHEA rely on σ1 receptors [410].

9.1.1 Promnesic and Anti-amnesic Effects of DHEA-S and Pregnenolone Sulfate

Sulfated neurosteroids enhance aspects of cognition whether administered systemically or into the CNS. Performance in a variety of cognitive tests is highly sensitive to pregnenolone sulfate and suggests roles in enhancing memory [409, 411–419]. A two-trial Y-maze memory test found that infusion of 5 ng of the steroid into the nucleus basalis magnocellularis promoted memory retention while higher levels (3 μg) increase performance regardless of whether it is administered prior to or after memory acquisition [414] (Fig. 9).

I.c.v. and intracisternal injection of DHEA or intracisternal DHEA-S at less than nmol levels improves long-term memory retention and, in the case of DHEA, counteracts dimethyl sulfoxide-induced amnesia [409, 420]. Both compounds enhance footshock active avoidance memory retention [421] and, along with pregnenolone, a separate measure of working memory in mice [419].

The physiologic source of the steroids is in part neural. Inhibition of P450scc by aminoglutethimide (Fig. 1b) and, hence, neurosteroidogenesis decreases postsynaptic AMPA receptor-dependent field excitatory postsynaptic potentials (EPSPs) in rat dentate gyrus granule cells [422]. This effect is rescued by pregnenolone, DHEA, or DHEA-S. Further testing with trilostane and letrozole (Fig. 1b) indicates that DHEA or DHEA-S along with estradiol may be the relevant endogenous steroids produced by these cells. Aminoglutethimide further inhibits NMDA receptor-dependent long-term potentiation (LTP) in this cell type, possibly by blocking pregnenolone sulfate synthesis [422].

Improvement in cognition involves cholinergic neurons and is stimulated in the brain by pregnenolone sulfate down to pmol levels [423]. Enhanced memory performance induced by pregnenolone sulfate correlates with increased Ach release in the neocortex and the hippocampus but not the striatum [411, 423, 424]. Similarly,

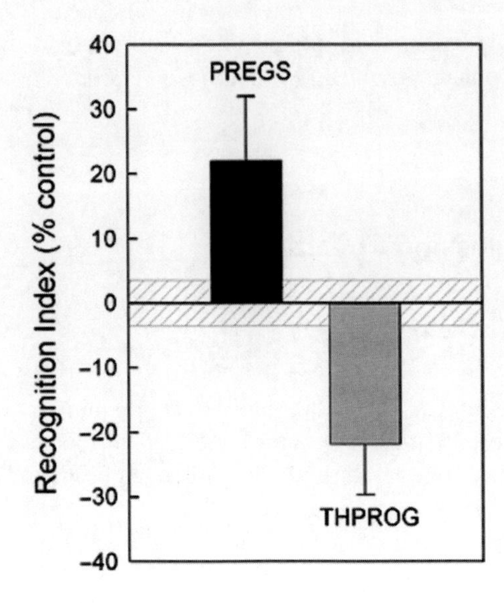

Fig. 9 *Opposing effects of pregnenolone sulfate (PREGS) and allopregnanolone (THPROG) on learning and memory.* A two-trial Y-maze test to measure recognition of novelty by male rats shows that recognition is enhanced by 5 ng of the former steroid infused into the nucleus basalis magnocellularis after memory acquisition not before as a percentage of control [414]. Infusion of 2-ng allopregnanolone prior to but not after the acquisition trial impairs performance. The range of SEM for control animals is indicated by the *striped box* (Figure reprinted with permission from [576])

DHEA-S stimulates hippocampal Ach release [425]. When steroid sulfatase activity is inhibited by long-term i.p. drug administration, plasma DHEA-S levels rise along with a corresponding increase in hippocampal Ach release and an easing of amnesia induced by the muscarinic AchR antagonist, scopolamine [426].

Pregnenolone sulfate however does not always improve cognition in studies. In passive avoidance testing, pregnenolone sulfate improves memory retention with 0.25-mA footshock over 2 s, but not with 0.3 when given s.c. [427] and worsens it at 0.5 with a 5-ng infusion into both hippocampi [315, 428].

Pregnenolone sulfate and DHEA-S promote these effects through positive modulation of NMDA and antagonism of $GABA_A$ receptors in the hippocampus and other relevant structures, as shown through studies that infuse steroids into the CNS [3, 423, 429–432]. These steroids reverse amnesia induced by NMDA receptor antagonists [429, 433]. On the other hand, the positive effects of pregnenolone sulfate on memory are lost in mice with a targeted knockout of the NR1 subunit in the CA1 [434].

The actions of both steroids may further involve σ1 receptors and glutamate release [410, 416, 435]. σ receptor antagonists disrupt potentiation by DHEA of excitatory responses by CA3 pyramidal neurons to NMDA [161] and the amelioration of scopolamine-induced learning and memory deficits by pregnenolone sulfate and DHEA-S [156, 415, 436]. In vivo infusion into the right ventricle with antisense oligonucleotides to σ1 receptor mRNA inhibits the rescue by DHEA-S of cognitive deficits induced by the NMDA receptor antagonist dizocilpine [437]. Antisense treatment does not prevent rescue by pregnenolone sulfate, indicating that the effects of the former steroid primarily utilize the σ1 receptor. At the same time, rescue of dizocilpine-induced memory impairments by a σ1 receptor agonist, DHEA-S, or pregnenolone sulfate is all opposed by progesterone with differing sensitivities [438]. Recent data further suggest that a primary effect of DHEA-S on hippocampal synaptic transmission in the rat is mediated through mGluR5 on AMPA receptors [439].

9.1.2 Estradiol

Generally, performance in select cognitive tests changes with the ovarian cycle. For instance, variations in serum steroids with cycle stage alter auditory fear extinction recall [440]. One ovarian steroid, estradiol, supports aspects of cognition, except at high doses which can impair spatial reference and working memory [441]. Ovarian sources of estrogen otherwise promote cognitive performance [442], such as in spatial and visual memory. Estrogen enhances key measures of memory formation and retention, like the amplitude of LTP in hippocampal neurons [194, 443]. This enhancement arises from estrogen's effects on the hippocampus and prefrontal cortex and its regulation of key neurotransmitters, like Ach [396].

While ovarian estrogen regulates hippocampal functions via efferents from subcortical regions like the median raphe [444–446], its direct effects may actually be mediated by estrogen produced in the CNS [447, 448]. Estrogen levels can greatly exceed that of the serum, suggesting a greater role for locally made steroid in hippocampal functions [9, 449]. One possibility is that ovarian estradiol-regulated fluctuations in GnRH during the cycle impact the hippocampus, where neuroestradiol synthesis is inducible by the GnRH receptor. Changes in receptor stimulation provide a putative mechanism for the observed cyclic changes in spine density in the hippocampus [450]. Inhibition of aromatase by letrozole increases GnRH receptor in hippocampal cultures, suggesting that the receptor is feedback regulated.

As discussed in part earlier, estrogen positively modulates synaptic size and plasticity in the hippocampus through ERα and ERβ receptors [451–455]. It affects the proliferation and differentiation of hippocampal neurons, maintaining dendritic spine densities in CA1 pyramidal cells, increasing spine densities in vitro, and inducing rapid declines in thorns in CA3 neuronal thorny excrescences in the rat hippocampus [456–458]. Inhibition of aromatase activity or knockdown of StAR expression in hippocampal neurons lowers neuroestradiol synthesis, causing alterations in spine densities, synaptic protein levels, and axon outgrowth [445, 459–461].

In CA1 pyramidal cells from adult female rats or E19–E20 rat embryos, estrogen causes phasic neuronal activation through suppression of $GABA_A$ receptor-dependent inhibition [462–465]. One mechanism for this is through (likely membrane) ERα stimulation of postsynaptic mGluR1a that in turn promotes retrograde endocannabinoid signaling and suppression of GABA release [466]. As remarked before, this hippocampal ERα/mGluR1a pathway is only active in the CA1 from female rats not males. Studies in adult and embryonic rodent neurons and slice cultures further reflect that increased excitability and glutamatergic transmission involve enhanced postsynaptic non-NMDA glutamate receptor activity, especially AMPA [467] and membrane ERα as well as ERβ-induced changes in AMPA receptor subunit composition [199, 468].

As noted previously, changes in dendritic spine density in CA1 pyramidal neurons involve NMDA receptors [469] and reduced $GABA_A$ neurotransmission [465]. These rapid effects also involve membrane ERα and ERK pathways [456, 470].

9.2 *Amnesic Neurosteroids*

Several neurosteroids oppose aspects of cognitive function. Pregnanolone can impair memory [419]. Pregnenolone sulfate also can have amnesic qualities as well. When 1.2-μM pregnenolone sulfate is injected into the lateral septum prior to training, performance by male rats in a novel object recognition test is completely disrupted [471].

9.2.1 Allopregnanolone

Allopregnanolone exerts opposite effects on basal forebrain cholinergic transmission to pregnenolone sulfate, selectively inhibiting Ach release in the hippocampus and prefrontal cortex [350] and worsening performance in spatial memory and two-trial learning and memory tasks [413, 414, 418]. Administration of the steroid i.v. to healthy middle-aged women impairs episodic memory, but not semantic or working memory [472]. In a Morris water maze study, the steroid i.v. more greatly impairs learning than memory when administered just prior to training [473]. A similar discovery was made in a two-trial Y-maze test in which the steroid was administered into the nucleus basalis magnocellularis [414] (Fig. 9).

Its effects on cognition likely stem from the potentiation of $GABA_A$ channels. Enhancement of spontaneous $GABA_A$ channel activity in hippocampal neurons by allopregnanolone blocks the increase in dendritic spine densities caused by estradiol in vitro [474]. Loss of the δ subunit improves acquisition of contextual freezing behavior in the conditioned fear test in females [475]. Males do not benefit from the loss, possibly reflecting the overriding influence of ovarian steroids that increase CNS allopregnanolone.

As discussed earlier, allopregnanolone differentially affects neurogenesis in the dentate gyrus depending on the dose and, by implication, cognition [476]. 3α-reduced steroids also correct impaired cognition in certain instances. Infusion i.v. of allopregnanolone or THDOC rescues cognitive deficits induced by dizocilpine in male rats [54]. While pregnenolone sulfate administration i.p. also rescues these deficits, inhibition of 5α-reductase activity blocks its effect as well as a concomitant rise in central allopregnanolone [53, 54]. These studies suggest that metabolism of pregnenolone sulfate to allopregnanolone, not pregnenolone sulfate itself, is important in this case. Furthermore, they provide examples of allopregnanolone increasing glutamatergic activity through an as yet undescribed mechanism.

Other factors govern the effects of allopregnanolone on memory. Pre-infusion of the steroid bilaterally into the CA3 hippocampus at a level that is antidepressant does not impair performance in a passive avoidance test [379]. Moreover, early life loss of 3αHSD activity in the neonate via finasteride injections results in impaired memory in this task later in life [315]. The impairment in the adult is incompletely addressed by infusion of allopregnanolone into the hippocampus, indicative of an important developmental role for 3α-reduced steroids.

As with anxiety, the role of allopregnanolone changes during puberty with the spike in α4β2δ $GABA_A$ receptors [477]. Increased levels of the receptor coincide with selective declines in cognition that are more apparent in females. LTP is not induced in the pubertal female CA1 hippocampus, but is in pubertal δ null mice, with receptor blockade or the addition of 30-nM allopregnanolone, the opposite of the steroid's effect on LTP prior to puberty. Moreover, i.p. allopregnanolone reverses learning impairments in pubertal mice. The beneficial effects of estrogen therapies on cognition may also involve changes in central allopregnanolone, as observed with DT56a [363]. Thus, gonadal status can determine the positive or negative role of allopregnanolone on learning and memory.

9.2.2 Progesterone

Progesterone reduces hippocampal LTP and synaptic transmission. It also opposes rescue of scopolamine-induced learning impairments by DHEA-S or pregnenolone sulfate as well as the rescue of dizocilpine-induced amnesia by a σ1 receptor agonist [436, 478]. The inhibitory actions of progesterone thus likely involve its antagonism of σ1 receptors. Its conversion to allopregnanolone may also play a role [442]. Indeed, increased levels of progesterone and allopregnanolone in humans are implicated in cognitive impairments during pregnancy [441]. On the other hand, progesterone supplementation s.c. in OVX rats improves many aspects of learning and memory performance [479].

9.3 Age-Related Memory Changes and Neurosteroid Rescue

With aging come declines in hippocampal pregnenolone sulfate and performance in the spatial memory tests, the Morris water maze and the Y-maze [480]. Old rats with the lowest levels of pregnenolone sulfate score the poorest in memory tasks. Hippocampal infusion of pregnenolone sulfate or a synthetic enantiomer transiently restores these cognitive deficits to youthful performance levels [480, 481]. Both pregnenolone sulfate and DHEA-S reduce age-related impairments in passive avoidance and a spatial memory plus-maze test in mice and for DHEA-S, active avoidance [433, 482]. These effects may be mediated through a nitric oxide-dependent pathway and PKC activation [433, 483].

Estradiol also improves learning and memory in aged animals of both sexes [484, 485]. Studies on male rat hippocampal slices suggest that estradiol reverses stress and age-related changes in LTP and long-term depression [442, 443]. Testosterone replacement also affords beneficial effects on cognition in men with low serum testosterone [390]. Its effects may partly involve aromatization to estrogen in the brain.

Mixed reports exist as to the disposition of allopregnanolone levels in the aging brain. Allopregnanolone reportedly declines with age in the cortex and pituitary by 16 and 12 months, respectively, while serum, testis, and, after 18 months, hypothalamic concentrations rise in male Wistar rats [486], whereas no change was discovered in neither it nor THDOC in the cerebral cortex of Sprague–Dawley rats by 18 months versus young adults [302]. However, changes in the former steroid with age may bear upon cognition.

Changes in circadian rhythm with aging also affect spatial memory [487]. In this case, allopregnanolone may also have a corrective role. Aging male rats with sleep-dependent spatial long-term memory deficits exhibit reduced levels of the steroid in the hypothalamus, pedunculopontine nucleus, and ventral striatum with no change in the hippocampus, despite possessing elevated serum pregnenolone [487]. This role may be indirect through its effects on neuroprotection or sleep patterns, as discussed later.

Studies on OVX rodents also link allopregnanolone and loss of ovarian function to cognition. Allopregnanolone or progesterone s.c. in OVX rats when given immediately after training improves performance in an object placement task, a test that recruits both the prefrontal cortex and hippocampus [488]. Allopregnanolone and its precursors improve other aspects of cognition in OVX rats. For instance, allopregnanolone reverses impaired performance in an inhibitory avoidance task elicited by a combination of estradiol and progesterone in OVX rats [489].

Continuous s.c. administration of progesterone starting at 6 months of age improves cognitive testing performance in 9–12-month-old mice [490]. A single s.c. dose of allopregnanolone also rescues cognitive deficits and promotes new neural cell survival in 15-month-old aged mice [491]. These long-term effects on cognition again may wholly derive from the neuroprotective quality of the steroids rather than a true promnesic action. Overall, the data indicate that allopregnanolone improves cognition in cases where normal CNS levels of the steroid are compromised. Steroid in excess of that can be deleterious.

Aged rats with spatial memory impairments also exhibit reduced hippocampal activity for cytochrome P450-7B1 (25-hydroxycholesterol 7α-hydroxylase), an enzyme that generates 7α-hydroxylated steroids [492]. One metabolite, 7α-hydroxypregnenolone, has been studied in vertebrates in regard to its melatonin-dependent synthesis in the brain and regulation of diurnal behavior [493]. Here, supplementation with the steroid improves spatial memory retention in impaired rats [492].

9.3.1 The Cognitive Effects of HRT in Women

In postmenopausal women, estradiol replacement improves aspects of cognition such as working memory [494–496]. As well, young to middle-aged women placed

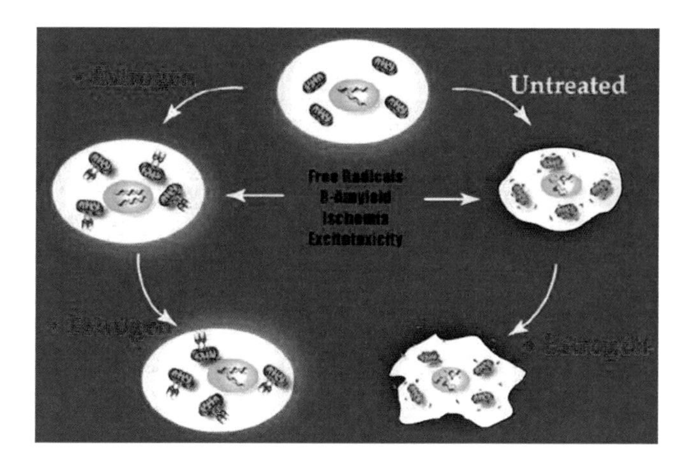

Fig. 10 *The healthy cell bias model for HRT.* Administration of estrogen prior to neurodegenerative insults with age and disease supports or maintains a healthy state for neurons. If neurons are left without estrogen, the cells will have changed such that exposure to estrogen after such insults now promotes their dysfunction and death (Adapted and reprinted with permission from [497])

immediately on HRT following oophorectomy or hysterectomy do not experience the cognitive decline seen in untreated patients [497]. Still, the beneficial effects of HRT on cognition wane with time after menopause [498]. On the other hand, administration of HRT to women 65 years and older participating in the Women's Health Initiative Memory Study (WHIMS) and other studies results in at best no cognitive improvement [498, 499]. These observations are the basis for the critical window hypothesis and the related healthy cell bias model (Fig. 10). These models hold that estrogen treatment is only beneficial if initiated at a time when neurons are still healthy (i.e., immediately or not long after the loss of normal ovarian function). Conversely, exposing unhealthy neurons to estrogen provides no benefit and can be detrimental.

A series of studies in a perimenopausal model, in which treated rhesus monkeys OVX at ~22 years old (aged) are compared to ~10-year-olds (young), points to the superiority of immediate cyclic estradiol administration over the constitutive estradiol typical of current HRT regimens to preserve cognitive function [455]. This research also finds that diminution of spine density in hippocampal neurons arises from both age and loss of estradiol. Hence, estradiol replacement does not solve all morphological age-related changes, but it does correct cognitive deficits.

Many studies in rodents support the basic findings in monkeys, with one exception. That study in OVX middle-aged rats finds no advantage, and it is suggested that the route of administration (oral vs. injected) and coadministration of progestin be in a manner that also simulates the cycle bear on the functional outcome [500]. Another significant consideration is that this study gave estrogen 3 days out of 4 to the rats, whereas other trials had less frequent dosing, to imitate peaks in estradiol (for instance, 1 out of 21 days in the monkey). A preliminary trial in the rat found that weekly administration of estradiol plus progesterone had a perceptible benefit over continuous dosing in a spatial memory task [501]. The dose of estradiol given can be important as well [502]. Additionally, inclusion of progestins can have negative or positive consequences (e.g., [502, 503]). In some studies, MPA is of more benefit than progesterone (e.g., in a small patient study [503]).

10 Neurosteroids and Reproduction

Reproduction requires the timed release of LH from pituitary gonadotropes. The gonadotropin stimulates gonadal steroid production, and its release is feedback regulated by these steroids. Gonadal steroids are required for sexual differentiation of the brain, reproductive functions, and sexual and gender-typical behavior, among other phenotypes. For example, testosterone shapes male sexual behavior by acting on target cells via AR and, after aromatization in the brain, ERs [504–507]. Neurosteroids also play key roles [508]. Neurosteroids are produced in relevant structures in the CNS, such as the amygdala, medial preoptic area (POA), and ventral medial hypothalamus (VMH) [21, 23, 24].

The effects of neurosteroids on reproduction are not always gender dependent. In rats bred for higher neonatal maternal separation anxiety, elevated serum and midbrain levels of progesterone and allopregnanolone correlate with heightened sexual activity in both adult males and females [509]. Potentiation of $GABA_A$ channel activity by 10-nM allopregnanolone inhibits GnRH release in male rat hypothalamic cultures [120]. Pregnenolone sulfate overcomes this suppression. Whether this occurs in the female is unclear. Other gender-specific differences are known.

10.1 Female Reproduction

Follicular development in the ovary culminates with a surge in LH release from the pituitary, an estrogen-dependent response that leads to ovulation. Ovarian estrogen and progesterone regulate the release of the gonadotropins follicle-stimulating hormone (FSH/follitropin) and LH indirectly through hypothalamic GnRH and directly at the anterior pituitary. Inhibitory feedback effects of gonadal progesterone on GnRH release rely at least in part on hypothalamic mPRα [510].

10.1.1 Progesterone and the LH Surge

Estrogen initiates the LH surge through progesterone and hypothalamic PR-A [511–513]. Studies on OVX/ADX animals among other observations lead to the conclusion that the source of progesterone is the hypothalamus [514]. Administration of supraphysiologic amounts (50 μg) of estrogen to OVX/ADX rats induces de novo progesterone synthesis in the medial basal hypothalamus and the LH surge through release of GnRH [515, 516]. Inhibition of hypothalamic progesterone synthesis blocks the LH surge [515].

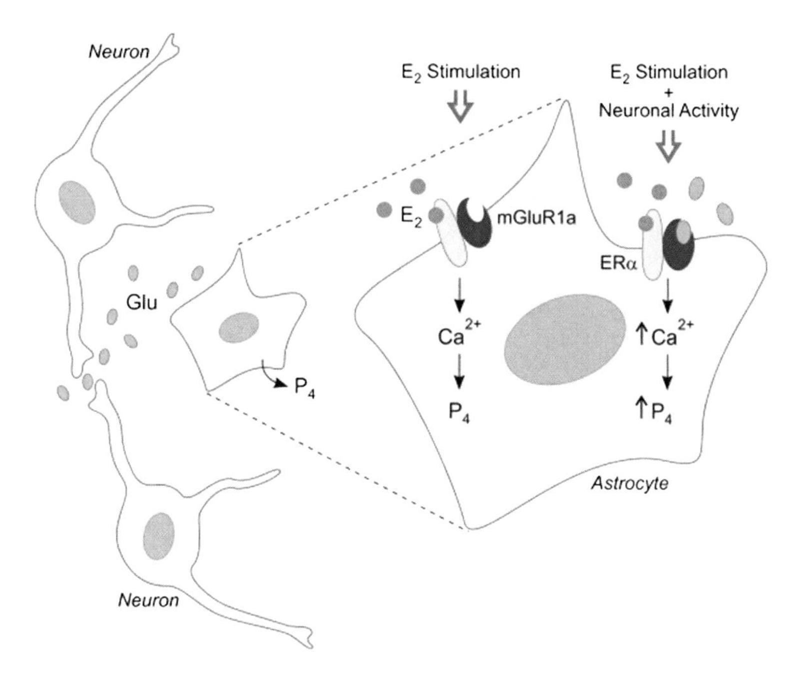

Fig. 11 *Proposed mechanism of how estradiol signaling and neuronal activity promote progesterone (P_4) synthesis in astrocytes.* Ovarian estradiol (E_2) binds to membrane ERα and activates the mGluR1a, increasing intracellular calcium. This increase in calcium results in steroid synthesis. Estradiol in combination with glutamate (*Glu*) secreted by neuronal terminals causes a larger surge intracellular calcium and a putative surge in progesterone that may be related to the LH surge (Adapted from Micevych et al. and reprinted with permission from [514])

Glia are the origin of this steroid. An exhaustive study that took into account ERα, ERβ, GPR30, and STX-sensitive mER-Gα_q concluded that estradiol elicits progesterone production through a membrane-anchored ERα associated with mGluR1a receptors [517]. Estradiol thus increases intracellular calcium, likely StAR activity through new synthesis or phosphorylation of the protein and steroidogenesis in hypothalamic astrocytes [67, 517–520] (Fig. 11). An intriguing question raised by this work is whether estrogen uses this pathway in other regions of the brain to also induce steroid production.

10.1.2 Allopregnanolone Effects on GnRH and Gonadotropins

Estrogen stimulates allopregnanolone levels in the hypothalamus and the anterior pituitary in aged and OVX rats [328–330]. Allopregnanolone in turn potentiates and DHEA-S inhibits the activation of postsynaptic $GABA_A$ receptors on hypothalamic GnRH neurons of diestrus female mice [521]. Through its effects on this receptor, allopregnanolone induces GnRH release down to 10 nM as indicated by murine GT1-1 immortalized neurons [129]. 6-μM allopregnanolone also upregulates NMDA-mediated release of the peptide hormone in hypothalamic explants from OVX rats primed to generate an ovulatory LH surge [200]. The facilitatory effect of allopregnanolone on GnRH release is site specific. Activation of hypothalamic $GABA_A$ receptors by i.c.v.-delivered allopregnanolone which provides access to the ventromedial nucleus (VMN) suppresses ovulation [382]. Separately, 5-μM pregnenolone sulfate potentiates NMDA receptor-mediated GnRH release by GT1-7 hypothalamic neurons [522]. In all cases, stimulated increases in GnRH release occur by 20–30 min. The data also suggest that age-related declines in hypothalamic allopregnanolone contribute to changes in ovarian function [328].

Allopregnanolone precursor and $GABA_A$ receptor modulator 3αHP is also produced in the anterior pituitary and inhibits basal and GnRH-stimulated FSH secretion [523–526]. This effect is gender neutral; is mediated through rapid, nongenomic effects on calcium and PKC signaling pathways utilized by GnRH; and is independent of metabolism to allopregnanolone [527, 528]. That said, a similar role for allopregnanolone is indicated by a small clinical study [529]. Women given i.v. allopregnanolone have reduced serum FSH and LH. Those administered an isomer that lacks $GABA_A$ channel activity do not experience this change. The study suggests that the high allopregnanolone levels observed in hypothalamic amenorrhea patients may contribute to the disorder through pituitary $GABA_A$ receptor activity.

10.2 Female Sexual Behavior

In mammals, proestrus and on into estrus comprise a period marked by rising serum estrogen and progesterone coincident with sexual receptivity and eliciting of mating posture or lordosis. Lordosis requires the action of ovarian estrogen on the hypothalamus. Activation of membrane ERα in VMN neurons and medial preoptic nuclear neurons in association with mGluR1a is crucial in expressing lordosis [530, 531]. Neurosteroids may also have important roles [508].

10.2.1 Progesterone

Estrogen can by itself elicit lordosis independent of PR activation [532]. On the other hand, proceptive solicitation behaviors rely on aminoglutethimide and trilostane-sensitive local progesterone synthesis and PR [532]. Estrogen increases PR numbers in the ventral portion of the VMN [533]. Progesterone utilizes these receptors to facilitate (though not cause) lordosis and appropriate tactile stimulation.

More is yet to be learned about this process with the list of steroid receptors expanding. In the absence of progesterone, estradiol stimulates PGRMC1 expression in and area relevant to female sexual behavior, the VMH [534]. On the other hand following estrogen priming, progesterone through PR represses this increase. As observed in female diestrus rats, PGRMC1 localizes in other hypothalamic structures as well relevant to sexual behavior like the amygdala, paraventricular nucleus (PVN), and the hippocampus. Interestingly, neuronal expression of PGRMC1 is absent in another key component for lordosis, the VTA, and it is increased in the hypothalamus of PR-null females but not males. These results point to a supporting role for PGRMC1 in female sexual behavior.

Incidentally, mPRα and mPRβ increase along with PR-B in the rat medial basal hypothalamus with proestrus [535]. The interplay of these various receptors in manifesting the effects of progesterone on sexual behavior remains to be elucidated.

10.2.2 Allopregnanolone

Levels of ovarian allopregnanolone and DHP peak in the brain during estrus in rats, associating with increased anxiolysis and social behavior that facilitate the presentation of receptive behaviors [536, 537]. Conversely, factors that increase stress or anxiety inhibit sexual behavior. Both progesterone and allopregnanolone can alleviate stress-induced inhibition through PR- and PR-independent mechanisms, respectively [321, 538, 539]. That said, MPA also inhibits solicitation behaviors in estrogen-treated macaques [339], and lowered libido is a reported side effect of the drug in women [540].

Allopregnanolone on the other hand may have a vital role in promoting sexual behavior beyond its PR-independent promotion of anxiolysis. Allopregnanolone levels rapidly increase in the rodent VTA with mating and, to a lesser degree, during proestrus with the onset of sexual receptivity [541, 542]. Moreover, paced mating selectively raises DHP and allopregnanolone in the midbrain, striatum, hippocampus, and cortex of estrus animals without any change in neither serum steroids nor CNS progesterone or estradiol [537, 543]. Only the midbrain experiences increases in DHP and allopregnanolone in diestrus mated animals [537]. In contrast to mating, an affective social interaction task does not alter levels of the steroids in these structures [543].

These data then suggest that local synthesis of allopregnanolone is specifically important for the manifestation of sexual behaviors. In fact, allopregnanolone or THDOC infused into the VTA enhances and maintains progesterone-facilitated lordosis in estrogen-primed OVX rats and hamsters [544, 545]. Inhibition of 3βHSD or 5α-reductase activity in the VTA reduces allopregnanolone midbrain levels and cells immunoreactive for the steroid in the VTA and attenuates lordosis [546, 547]. Infusion of allopregnanolone into the VTA rescues this behavior [546].

The effect of allopregnanolone is site specific. Delivery of allopregnanolone by i.c.v. in OVX rats suppresses sexual behavior possibly through effects on the VMN [382, 548]. Circulating allopregnanolone may similarly access the VMN and be inhibitory. Allopregnanolone levels in the VMH but not the cortex fluctuate with the ovarian cycle, with its lowest levels reached at proestrus [382]. Administration of antiserum to allopregnanolone i.c.v. during proestrus augments lordosis [382].

Thus, female sexual behavior may be a product of facilitation by progesterone and disinhibition by allopregnanolone in the VMN, while in the VTA, allopregnanolone augments the activity of the VMN.

At the cellular level, the VTA expresses little PR. Neuronally derived progesterone is instead metabolized to allopregnanolone and subsequently acts on $GABA_A$ receptors with effects on D1 receptors and cAMP pathways to promote sexual responsiveness, including appetitive and consummatory behaviors [257, 258, 541, 542, 549]. Similar to allopregnanolone, infusion of androstanediol produces paradoxical effects on lordosis and promotes aggression in female rats, possibly through $GABA_A$ receptor inhibition in the VMN and medial POA [550–552]. More is yet to be learned regarding the role of this steroid in female sexual behavior.

In certain cases, allopregnanolone is sufficient for lordosis independent of PR activation. Systemic administration of high doses of allopregnanolone to estrogen-primed PR-knockout OVX mice induces lordosis in middle-aged and aged females [553]. This suggests a strategy whereby the steroid can rescue receptivity when PR levels in the VMN are compromised [328]. Moreover, infusion of the steroid into the midbrain central gray, which accepts projections from the VMN as part of the lordosis circuit, also facilitates lordosis likely through $GABA_A$ receptors in estradiol-primed OVX rats [554].

10.2.3 The Role of Steroids in Female Libido

Little is known about the role of neurosteroids in human libido. Aside from the impact of physical symptoms and unlike in men, sexual behavior in women is not clearly reliant upon gonadal status. Psychological factors like societal influences or depression have a larger role [555]. Still, gonadal status has a role since mood disorders like depression correlate with changes in serum steroids [11]. Anxiolytic steroids and neurosteroids may be facilitatory. Indeed, a recent small clinical trial with early menopausal women found low-dose DHEA to be as effective as HRT in increasing the frequency of sexual intercourse [556]. That said, SSRIs (and by implication, central allopregnanolone) are generally known to oppose sexual interest in both sexes. However, postmenopausal women given paroxetine see improved sexual interest [391].

10.3 Male Reproduction

In the male, spermatogenesis depends on LH and FSH to stimulate testosterone synthesis and support sperm maturation. Progesterone, testosterone, and estradiol increase FSH secretion from rat anterior pituitary cells, while estradiol and DHT suppress gonadotropin release in prepubertal male rats [523, 525].

Gonadal testosterone feedback inhibits pituitary LH release. Loss of the GPCR6A receptor allows exogenous testosterone to induce a fivefold increase in serum LH

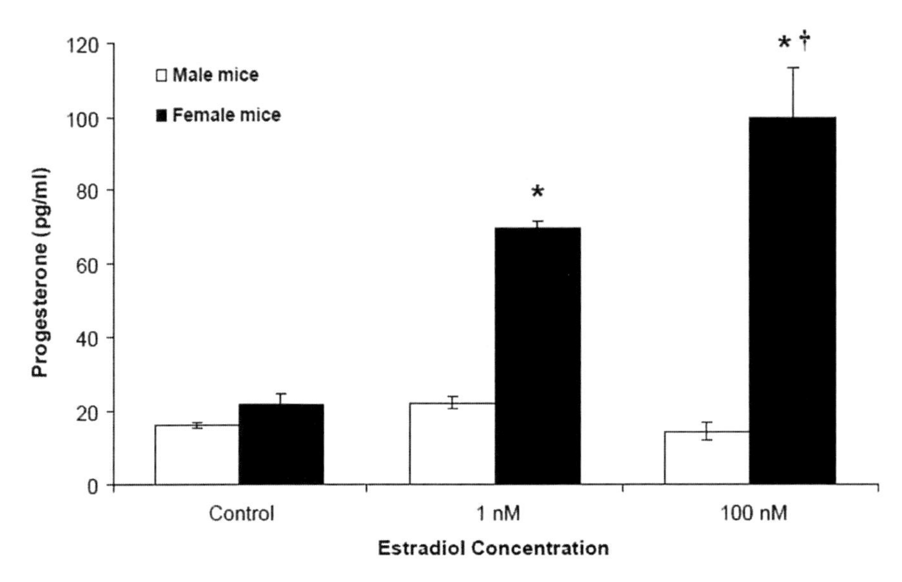

Fig. 12 *Estradiol inspires progesterone production in postpubertal female but not male hypothalamic astrocytes.* *Significantly different from female control and all male groups. †Significantly different from female astrocytes stimulated with 1-nM estradiol (Adapted and reprinted with permission from [557])

instead of suppressing the gonadotropin release [231]. Thus, gonadal testosterone relies on a nongenomic pathway to maintain appropriate LH secretory patterns.

Unlike in females, estrogen neither stimulates hypothalamic progesterone production in GDX/ADX males [515] nor by astrocytes isolated from the male hypothalamus [557] (Fig. 12). Micevych and colleagues hypothesize that the reason lies in the muted elevation in intracellular calcium with membrane ERα activation [514]. In contrast, repeated estradiol exposure leads to increased responsiveness by hypothalamic astrocytes in females ("positive feedback"), culminating in a sufficient calcium response to induce progesterone synthesis and the LH surge. This feedback mechanism is lacking in the male, meaning that select astrocytes are sexually differentiated. Indeed, glia within the developing hypothalamus exhibit a sexually dimorphic response to testosterone in terms of maturation and differentiation [558, 559] and possess lower levels of ERα [286]. Moreover, the ERα-mGluR1a interaction may simply be missing, given its gender dependence in other neural cell types.

10.4 Male Sexual Behavior

Through the medial POA and the amygdala, estradiol rapidly facilitates copulatory behavior in several animal models [560–562]. Both gonadal androgen and estrogen converted from testosterone in the brain are further critical for sexual behavior and the maintenance of sensitivity for sexual stimulation.

Neurosteroid regulation of $GABA_A$ receptors also influences male sexual behavior. Allopregnanolone and pregnenolone sulfate modulate neuronal activity in the medial POA [563–565]. Locally synthesized progesterone may be involved. Loss of PR or s.c. infusion of the antiprogestin RU486 (mifepristone) enhances mount and intromission frequency in the absence of changes in testosterone or testicular function or significant reductions in anxiety [566]. At the same time, AR expression is elevated in the medial POA and bed nucleus of the stria terminalis in PR-null mice.

Neurosteroids and steroidogenic enzymes are also found in components of the vomeronasal pathway, important for relaying stimulatory signals to the POA and bed nucleus of the stria terminalis to stimulate sexually experienced male rodents [10, 21, 24]. Preference for the odors of estrus females is enhanced by i.c.v. administration of 3αHP and reduced by pregnenolone sulfate [567, 568].

10.5 Male-Typical Behavior

Neurosteroids potentially influence gender-typical behaviors like aggression. Chronic treatment with DHEA inhibits aggressiveness by GDX male mice toward lactating female intruders [346]. This behavior is linked to GABAergic tone and a decline in central pregnenolone sulfate. Trilostane s.c. in castrate males elevates CNS pregnenolone but not pregnenolone sulfate and reduces belligerence [569]. The effect is not additive with DHEA. The change in behavior with trilostane may neither simply reflect a loss of progesterone, since ablation of PR only decreases aggressiveness toward infants not adult males [570]. Notably in the former study, anxiolytic allopregnanolone levels slightly increase with either trilostane or DHEA, and trilostane also increases THDOC, all in spite of its systemic inhibition of 3βHSD [569]. These data infer that the drug has low penetrance into the brain to inhibit neurosteroid synthesis or is used at a submaximally effective dose.

Similar inhibitory effects on behavior and pregnenolone sulfate by DHEA are reported for androgenized females [417]. Androstanediol self-administration also reduces aggressiveness and may be rewarding in male hamsters [571]. Thus, 3α-reduced steroids in addition to DHEA regulate male aggressive behavior. The source of these steroids in vivo may be the CNS. Several other studies using a social isolation stress model also support this notion.

Social isolation stress selectively reduces allopregnanolone and 5α-reductase levels in corticolimbic structures in males, leading to increased aggressive behavior in a resident-intruder test [317, 318, 572, 573]. SSRIs rescue the decline in allopregnanolone and change in aggression [573, 574]. The increase in aggression is androgen dependent, only observed at high levels in castrate males and intact and OVX females chronically treated with testosterone [573]. Social isolation stress by itself neither alters allopregnanolone levels in an important structure in aggressive behavior, the olfactory bulb, for castrate males nor intact and OVX females. This result indicates a direct effect of exogenous testosterone on local allopregnanolone synthesis. In a similar experiment, pregnanolone or S-norfluoxetine infusion at a dose that does not alter 5-HT reuptake into the basolateral amygdala reduces aggression to a same-sex intruder, with the latter drug also increases allopregnanolone in the amygdala and hippocampus [575]. Infusion of S-norfluoxetine into the striatum on the other hand affects neither resident allopregnanolone levels nor aggression. Thus, allopregnanolone synthesis in select regions of the brain may regulate male-typical aggression.

11 Sleep, Anesthesia, and Hypnotic Effects

The opposing actions of allopregnanolone and pregnenolone sulfate on cholinergic neurons affect sleep [576]. Infusion of pregnenolone sulfate by i.p. or into the nucleus basalis magnocellularis (NBM) or pedunculopontine tegmentum nucleus (PPT) increases the duration of paradoxical or rapid eye movement (REM) sleep without affecting slow-wave non-REM sleep and wakefulness [576–579]. This correlates with improvements in spatial memory performance [578]. A higher dose of the steroid in the PPT (10 vs. 5 ng) additionally increases slow-wave sleep and adversely affects parameters of wakefulness, such as vigilance [577]. The mechanism of pregnenolone sulfate action involves Ach release elicited through its modulation of NMDA and $GABA_A$ receptor activities in the frontal cortex and amygdala [423, 424, 577]. Its effects at increased doses may include kainate or AMPA receptors.

Progesterone and allopregnanolone dose dependently decrease the onset of non-REM sleep, decrease REM sleep, and alter other parameters of sleep similar to $GABA_A$ receptor agonists [580–582]. Administration of allopregnanolone, which inhibits Ach release, by i.p. or into the NBM but not the PPT decreases the duration of paradoxical sleep [350, 576, 578]. This effect is opposed by picrotoxin [583].

Old male rats with reduced circadian locomotor activity exhibit impaired sleep-dependent spatial long-term memory and low allopregnanolone in selective areas of the CNS, including the hypothalamus [487]. Oral indomethacin reproduces this reduction in activity in young rats and correlates with a specific loss of allopregnanolone in the hypothalamus, suggesting a peripheral origin for the steroid [487]. The effects of allopregnanolone involve δ subunit-containing $GABA_A$ receptors. Sleep time induced by a synthetic analog of the steroid or by pregnanolone is reduced in δ null mice [313].

Another 3α-reduced steroid THDOC induces sleep and increases non-REM sleep in rats [584]. Pregnenolone also enhances delta activity in non-REM [585]. As well, DHEA and DHEA-S alter REM or non-REM sleep [586, 587].

Sex steroids originating in the periphery unquestionably impact sleep. Estradiol affects sleep/wake states and studies in castrate rats as in the clinic correlate it and progesterone's effects to changing gonadal supply of the steroids, such as occurs with aging [588, 589].

Steroids like progesterone and testosterone are also potent anesthetics [590]. The anesthetic effect of progesterone relies on conversion to allopregnanolone with subsequent action on $GABA_A$ receptors. Anesthetic activity elicited by either steroid is preserved in PR-knockout mice, and progesterone's effect is blocked by finasteride [591]. I.c.v. PEA augments the hypnotic effect induced by the barbiturate pentobarbital through increases in StAR, P450scc, and subsequent allopregnanolone synthesis in the brainstem [77]. Therefore, current evidence implicates both local and peripheral steroids in sleep and hypnotic and anesthetic effects.

12 Other Functions

Neurosteroids are implicated in other CNS functions. Allopregnanolone increases $GABA_A$ receptor-mediated suppression of respiratory frequency in perinatal rats, while DHEA-S opposes it [592]. Allopregnanolone selectively induces vasopressin release from the posterior lobe of the pituitary in vitro through a $GABA_A$ receptor-mediated mechanism [593]. Neurosteroids synthesized in the retina may regulate $GABA_A$ receptors in neurosynaptosomes, suggesting a role in information processing [594].

Neurosteroids have a role in auditory processing pathways [595]. Toxin-induced elimination of the primary inhibitory GABAergic input from the contralateral dorsal nucleus of the lateral lemniscus into the central nucleus of the inferior colliculus of anesthetized rats increases neuronal excitability. This increased excitability starts to vanish by 20 min later, indicating that the remaining inhibitory inputs are potentiated to compensate. This change is blocked by i.p. finasteride. Immunohistochemistry of frozen sections confirms that the lesion selectively promotes a finasteride-sensitive elevation in the number of cells producing allopregnanolone in the inferior colliculus. Thus, increased neuronal allopregnanolone synthesis leads to the promotion of compensatory $GABA_A$ receptor currents and suggest an integral role for neurosteroidogenesis in auditory pathways.

13 Neurosteroids in Disease and Injury

13.1 Neuroprotection in the Brain

A large number of studies firmly establish the protective effects of neurosteroids day to day and under and as a prophylactic against many pathologic conditions. Neurosteroid synthetic machinery and synthesis correspondingly change in response to injury and with aging. Central StAR levels rise with aging in the rat [24]. Rat hippocampal production of StAR and neurosteroids increase in response to excitotoxic stimuli like NMDA and kainate [24, 88], and rat inferior olivary nuclear levels of StAR and P450scc mRNA rise in response to 3-acetylpyridine [69]. Degeneration of the inferior olivary nucleus by the toxin also selectively elevates StAR mRNA in the cerebellum [69]. Asphyxia provokes increased cerebrocortical P450scc and 5α-reductase II synthesis along with CSF allopregnanolone in fetal sheep [596]. Allopregnanolone levels also change in the brain independent of the serum in young lambs exposed to hypoxia and LPS [597].

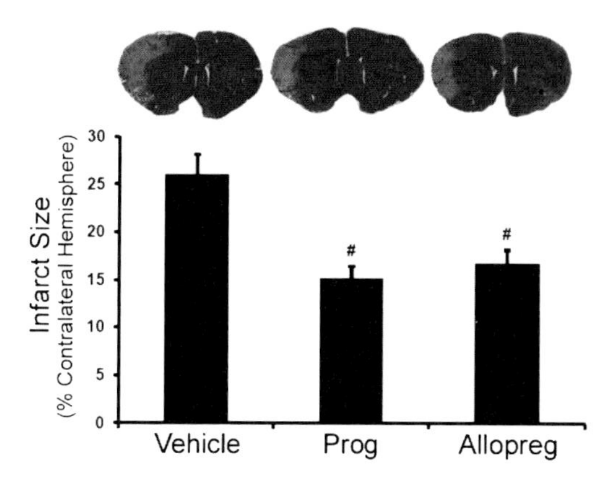

Fig. 13 *Progesterone (Prog) and allopregnanolone (Allopreg) reduce infarct size in rats with permanent MCA occlusion.* Infarct regions are visualized as lighter areas in cresyl violet-stained coronal brain sections 72 h post-occlusion. The regions are quantified in the histogram as compared to the area of the contralateral side for all groups. The steroids significantly reduced infarct size by 35–41% ($p < 0.05$) (Adapted and reprinted with permission from [604])

As part of a neuroprotective response, newly synthesized neurosteroids increase cell-survival factors and brain-derived neurotrophic growth factor (BDNF) and dampen the inflammatory response, among other effects. For instance, estradiol and progesterone inhibit the proliferation of microglia as shown in culture [598]. Allopregnanolone and progesterone both lower the level of inflammation, neurodegeneration, and apoptosis as well as cognitive impairments in cortical contusion models [599–601]. Paroxetine administration increases plasma BDNF in postmenopausal patients and may also positively affect aging related processes in the CNS at least partly through neuroallopregnanolone [391].

13.1.1 Progesterone and Allopregnanolone

Progesterone reduces cell damage and improves outcomes of focal ischemia induced by occlusion of the middle cerebral artery (MCA), a classic stroke model [602]. The neuroprotective effect of progesterone is mediated in part through modulation of the inflammatory response, including limiting permeability of the blood–brain barrier, infarct size, and edema [602–604] (Fig. 13). Insight into this mechanism is gained from the observation that i.p. progesterone does not reduce edema volume caused by an occluded MCA in NOS-2 null mice [605].

Two clinical trials focused on treating traumatic brain injury (TBI) find that i.v. progesterone dramatically reduces morbidity and mortality and improves functional outcomes without serious side effects [606]. Several studies are now in phase III trials and another for pediatric TBI is in phase I/II.

Progesterone's anti-apoptotic effect in an ovarian cell type is exclusively mediated through PGRMC1 [172]. As it happens, PGRMC1 is upregulated in TBI

[173]. Therefore, progesterone's effects could be mediated through PGRMC1, aside from any metabolism to neuroprotective allopregnanolone. Further evidence of a direct effect by progesterone comes from studies with non-metabolizable synthetic progestins. For example, Norgestrel protects photoreceptor cells and decreases apoptosis in models of retinal degeneration [607]. *The proof comes from the use of PR-null mice and a synthetic progestin that can't be metabolized to allopregnanolone. They further confirm that allopregnanolone operates through a non-PR-dependent pathway to afford neuroprotection in this model.

Like progesterone, allopregnanolone is protective against focal ischemia and potentially more potent [604] (Fig. 13). Allopregnanolone typically affords protection through $GABA_A$ potentiation [608]. The steroid blocks picrotoxin- and kainate-induced cell death in cultured hippocampal neurons [282, 608, 609]. However, allopregnanolone is ineffective against cell death in anoxic rat E18 cerebral cortical cultures [610]. Stimulation of $GABA_A$ receptor activity as the steroid encourages is neurotoxic for P3 but not P10 immature Purkinje cells in rat cerebellar sections [611]. The reason for this difference relates to these ages being a time when receptor activation switches from being excitatory to inhibitory. In both P3 and P10 sections, RU486 is protective. Often acting as PR antagonist, it has been speculated that RU486 exerts this effect through PGRMC1, not PR [280, 612].

Progesterone, DHP, and allopregnanolone also partially reverse age-related changes in myelination in the rat [613]. Progesterone s.c. promotes caudal cerebellar peduncular neuron remyelination by oligodendrocytes following ethidium bromide-induced lesions [614]. This enhancement is age dependent, occurring in 9-month-old but not 10-week-old male rats.

Both progesterone and allopregnanolone also increase an index of myelination, myelin basic protein, in cerebellar cultures of P7 rats and mice [615]. Progesterone supports this partly through the proliferation and maturation of oligodendrocyte precursors [616]. Mechanistically, progesterone accomplishes this through $GABA_A$ and PR [615]. Inhibition of 5α-reductase diminishes progesterone's effects on myelination, while inhibition of the $GABA_A$ receptor blocks allopregnanolone's influence. However, progesterone has no effect on myelination in cerebellar slice cultures from P7 PR-null mice or in the presence of RU486 in P7 rat cerebellar preparations, while a PR agonist has similar efficacy to progesterone. Thus, the bulk of progesterone's pro-myelinating actions requires PR.

13.1.2 Pregnenolone Sulfate

Pregnenolone sulfate sensitizes hippocampal neurons to chronic NMDA exposure and thereby increases cell death [617]. Both pregnenolone sulfate and DHEA-S also increase kainate-induced toxicity [87]. Pregnenolone sulfate further induces caspase-mediated excitotoxic retinal cell death through NMDA receptors [618, 619]. Interestingly, NMDA acting through NMDA and extrasynaptic $GABA_A$ receptors also stimulates pregnenolone and/or pregnenolone sulfate production in the retina [87]. Blockade of steroid formation reduces cell death induced by NMDA.

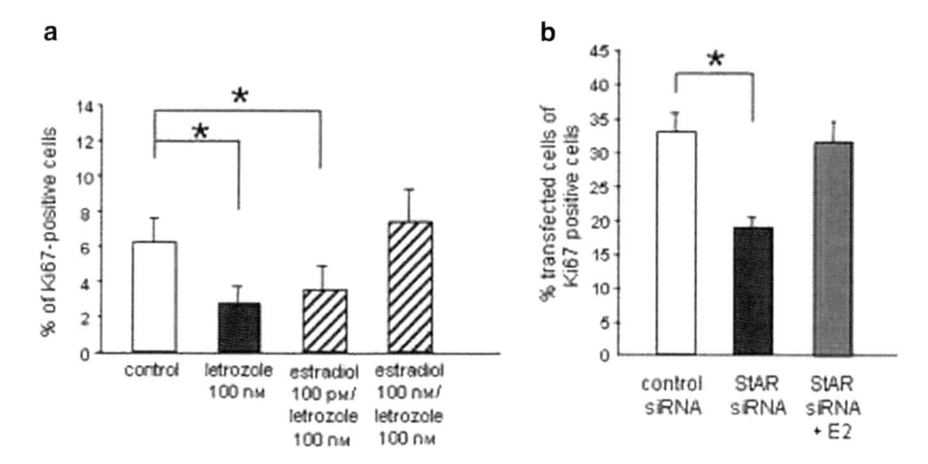

Fig. 14 *Proliferation of hippocampal granule cells is facilitated by* de novo *synthesis of estradiol*. (**a**) Blocking estradiol synthesis with letrozole (**b**) or siRNA knockdown of StAR reduces cell proliferation in vitro as judged by Ki67 labeling. In both cases, addition of estradiol (E2) to the media rescues this loss (Adapted and reprinted with permission from [461])

Pregnenolone sulfate can be neuroprotective. Consistent with its inhibitory effect on AMPA receptors, the steroid blocks AMPA neurotoxicity in cortical sections [620]. At the same time, pregnenolone fails to significantly protect against AMPA toxicity. The protective effects of pregnenolone in other situations aside from microtubule integrity as mentioned earlier may derive its metabolism to estrogen (reviewed in [621]).

13.1.3 Estrogen

Estrogen protects against a variety of stressors and toxic insults including anoxia, kainate, glutamate, NMDA, and AMPA [195, 622–627]. For instance, estradiol preserves retinal ganglion cells following optic nerve axotomy in OVX rats, indicating that retinal estradiol production has a protective function [628, 629]. Some regional cell loss occurs in this model with ovariectomy, supporting a role for gonadal estrogens as well [629]. Estradiol along with pregnanolone sulfate opposes pregnenolone sulfate-induced apoptosis in rat retinal cells in vitro [618].

Neuroprotection by estradiol arises from a combination of long-term and acute effects mediated through ERβ blockade of pro-apoptotic gene expression, increased BDNF, and G-coupled membrane receptors that activate anti-apoptotic pathways [623, 630–633]. Protective pathways may also be stimulated through the activation of L-type VGCCs [634]. One study using a spinal cord neurodegeneration model supports the notion that activation of ERα is both anti-inflammatory and neuroprotective, whereas ERβ only elicits the latter response [635].

Neocortical levels of ER-X dramatically rise in response to MCA occlusion, suggesting a supporting role for this putative estrogen receptor [213]. A further rise in central allopregnanolone elicited by estradiol may contribute to estrogen's protective effects. The increase partly or wholly corrects regional deficits in the former steroid due to aging or ovariectomy in female rats [328–330].

Many studies focus on the neuroprotective role of estrogen specifically in the hippocampus. Estrogen preserves cognitive functions against disease, aging, and stroke [636], mitigating cell death and stimulating neuronal proliferation in the hippocampus and other structures. Knockout of aromatase does not increase the rate of loss of hippocampal neurons, but does increase their susceptibility to neurodegeneration by various toxic insults [621]. The ovary is a key source of neuroprotective estrogen. Ovariectomy decreases spine density in CA1 pyramidal cells, while s.c. estrogen blocks this decline [457].

Neuronal sources of estrogen in the hippocampus and other brain regions are also important [447, 449]. Neuronal estrogen preserves spine density and presynaptic bouton number in hippocampal slice cultures [637]. Its loss with letrozole reduces these morphological markers, increases apoptosis, and decreases neuronal proliferation [461, 637] (Fig. 14). While estradiol can rescue the effects of letrozole, in control cultures the addition of estradiol in excess of ambient levels to simulate a substantial peripheral contribution fails to improve these parameters. Thus, ongoing estradiol synthesis by hippocampal neurons is more relevant and sufficient for estrogen-modulated synaptogenesis and cell survival, whereas ovarian estrogen may support hippocampal neurons indirectly, as remarked earlier [445]. Given that hippocampal neuroestrogen changes with the ovarian cycle [9], loss of ovarian function with age impacts not only serum estradiol but endogenous sources of the steroid and, thereby, hippocampal function.

Changes in estradiol in response to injury likely arise from de novo synthesis, not peripheral precursors. Brain injury increases aromatase expression and activity [621, 638]. Damage to the inferior olive by i.p. injection of 3-acetylpirydine increases StAR, P450scc, and aromatase mRNA in cerebellar neurons, suggesting a rise in estradiol synthesis from cholesterol [69]. Inhibition of aromatase activity in GDX mice increases vulnerability to neurotoxins [621, 639]. Estradiol protection against permanent cerebral ischemic injury in OVX mice is mediated by ERα and is observed after 8–16 h when given s.c. [640, 641] (Fig. 15).

Interestingly, while aromatase primarily localizes to neurons prior to injury, the important neuroprotective increase in estradiol post-injury may actually result from the gain in aromatase activity in reactive astrocytes [642]. Indeed, kainate injected i.c.v. reduces P450scc expression in neurons while increasing it in astrocytes in the hippocampus [643].

As mentioned previously, estradiol is dose dependently protective or pro-apoptotic for astrocytes [286, 287]. Estradiol utilizes ERα to reduce astrocyte death with toxic insults like ischemia [287]. Estradiol also protects endothelial cells in the brain [644].

13.1.4 Testosterone

In males, neuroprotection by testosterone is partly due to its conversion to estrogen in the CNS [621]. As well, physiologic levels (in this case, reckoned as 4 nM) of

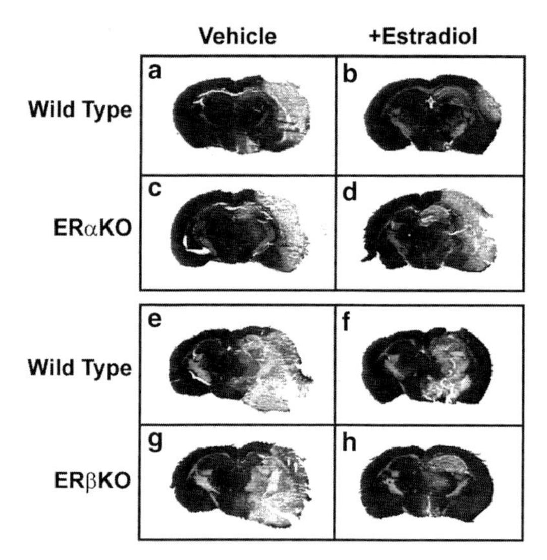

Fig. 15 *Estradiol reduces infarct size in mice with permanent cerebral artery occlusion through ERα.* In coronal brain sections stained with hematoxylin and eosin from oil vehicle-treated OVX ERαKO, ERβKO, and wild-type littermates following permanent cerebral ischemia in the right hemisphere, extensive brain injury is observed (lighter areas) (**a**, **c**, **e**, and **g**). Wild-type and ERβKO OVX mice given estradiol replacement realize reduced infarct sizes (**b**, **f**, and **h**). However, the condition of hormone-treated ERαKO mice is unchanged (**d**), indicating that estradiol utilizes an ERα-dependent mechanism to exert neuroprotection (Adapted and reprinted with permission from [640], copyright (2001) National Academy of Sciences, USA)

testosterone protect primary human fetal neurons against serum-deprivation-induced cell death through ARs [645]. Higher levels of the steroid (1 μM) protect rat cerebellar neurons against oxidative stress, also through AR [646]. In human SH-SY5Y neuroblastoma cells, 100-nM testosterone rapidly induces neurite outgrowth through a calcium-dependent mechanism, apparently through mARs (GPCR6A) and AR [647]. However, the effects of testosterone are biphasic. Higher levels (1–10 μM) induce apoptosis through a similar mAR-type pathway [648]. Similar results are obtained in mouse hippocampal HT-22 cells, where the androgen enhances glutamate toxicity at 10 μM while 10-μM estradiol rescues the cells [649]. In the MCA occlusion stroke model, testosterone and DHT reduce at low or increase at high levels of supplementation s.c. infarct size in GDX male rodents [649, 650].

13.1.5 Protective Effects of HRT

Age-matched women are less likely to suffer from ischemic stroke than men [651]. However, women with declining levels of sex steroids due to menopause see their risk of stroke quickly rise [651]. While estradiol is effective in experimental stroke models, HRT studies predict estradiol supplementation increases the risk of ischemic stroke and poor prognosis [651]. This leaves the question open as to how to capture the neuroprotective effects of estrogen. Timing of administration may be key for an optimal therapeutic outcome, given the critical window hypothesis and data on cyclic administration of HRT.

As noted, a majority of laboratory studies suggest that progesterone-replacement therapy may be beneficial prior to or following stroke in both sexes for morphological and functional outcomes [651]. Allopregnanolone may be even more beneficial. However, the long-term prophylactic effects of progesterone are another question. The inclusion of progestins in estrogen replacement regimens blocks the neuroprotective effect of estrogen [652, 653]. Progesterone opposes, for instance, estradiol-induced increases in ERβ and downstream BDNF against NMDA toxicity [633]. Studies with AD mice as detailed later indicate that a major problem is also timing – when and how often progesterone is administered [654, 655].

Issues have also arisen with HRT for postmenopausal women with MPA. The drug antagonizes neuroprotection by estrogen, interferes with neuroprotective allopregnanolone production, and poorly binds putatively neuroprotective mPRα [224, 656–658]. Unlike progesterone, MPA is not metabolizable to allopregnanolone and competitively inhibits allopregnanolone synthesis [659, 660]. However, the data are not unanimous on these points. Ovariectomy in rats reduces allopregnanolone in the brain [331], which as noted before, estradiol can restore. One study found that oral 14-day administration of up to 0.2 mg/kg of MPA to OVX rats does not reduce CNS allopregnanolone nor interfere with estradiol's restoration of allopregnanolone levels [661].

13.1.6 DHEA and DHEA-S

Timing is also an issue with DHEA. In a rat model of transient global cerebral ischemia, a single i.p. dose is strongly neuroprotective if given 3–48-h post-occlusion or worsens neuronal death and performance in the Morris water maze if given within 1-h pre- or post-occlusion [662]. The exacerbation of ischemic and reperfusion injuries is potentially mediated via σ1 and NMDA receptors, while DHEA's neuroprotective effects involve just the former receptor. Its sulfate conjugate is similarly neurotoxic in this model.

As well, 10–100-nM DHEA protects rat E18 hippocampal neurons in culture against NMDA, AMPA, and kainate toxicity, but is most effective against at least NMDA as a pretreatment [663]. DHEA-S also protects against NMDA toxicity, but only as a 6-h pretreatment and at a higher concentration than DHEA (100 nM vs. 10 nM) [663]. A second NMDA study on rat E19 hippocampal neurons found that the minimum concentrations for DHEA and estradiol were far greater than the 100 nM for DHEA-S and that co-treatment was the most important factor for inhibition of cell death [634].

Preincubation of rat E18 cerebrocortical neurons for 24 h with either 0.01–1-μM DHEA or 1-μM DHEA-S increases survival after 2-h anoxia [610]. However, supplementation with 0.1-nM DHEA-S is pro-apoptotic [610]. The effects of these steroids are independent of metabolism to estradiol, and they do not alter cell survival in the absence of insult [610]. DHEA-S and, less effectively, DHEA (maximally at 100 vs. 10 nM) also increase neuronal and glial survival and/or differentiation in cultures from E14 mice [420].

Pretreatment with DHEA s.c. for 5 day opposes the generation of lesions caused by the introduction of 5–10-nmol NMDA into the adult rat CA1/2 hippocampus [663]. In E14 mouse brain cell cultures, 100-nM DHEA and 10-nM DHEA-S maximally promote neuronal survival and reduce astroglial proliferation [289]. The steroids further promote astrocytic differentiation or survival. Neuroprotection in the hippocampus may also be extended by local synthesis of 7α- and 7β-hydroxylated forms of DHEA and pregnenolone [664].

Additionally, DHEA protects against oxidative damage to hippocampal neurons and glutamate toxicity [665, 666]. These effects involve its inhibition of nitric oxide synthase [634] and G protein-coupled receptors that reduce cell sensitivity to glucocorticoids and activate cell-survival pathways [666–669]. Receptor activation is antagonized by androgens and glucocorticoids [668].

The neuroprotective effects of DHEA-S against NMDA excitotoxicity rely at least in part on σ1 receptor activity [634]. Progesterone antagonizes σ1 receptor-dependent DHEA-S neuroprotection of ischemia-induced impairment in LTP but has no protective effect itself [670]. The protective effects of DHEA and DHEA-S for embryonic neurons may also be exerted via antagonism of $GABA_A$ receptors, which in the developing CNS are generally excitatory. Additional neuroprotection by DHEA in general as suggested before may arise from its metabolism to estrogen [621].

DHEA-S is not universally protective against toxic insults and can promote cell loss. The sulfate decreases levels of pro-survival kinase Akt and increases apoptosis in neural precursor cells from rat embryonic forebrain at the same 50–100-nM level at which DHEA promotes survival in this culture system [669].

13.2 Neurosteroids in Neurodegenerative Disorders in the Brain

Declines in CNS steroids correlate with the progress of neurodegenerative diseases such as Parkinson's disease [671] and Niemann-Pick disease type C (NPC) [672]. Repletion of these steroids therefore represents a potential therapeutic strategy. For instance, 500-nM progesterone and 250-nM allopregnanolone reduce aggregation of mutant Q74-repeat huntingtin protein in transfected astrocytes without inducing cell death [673]. The mechanism involves RU486-sensitive PR. Estradiol s.c. also protects against the cuprizone mouse model of demyelinating diseases, reducing measures of inflammation, forestalling increases in microglia, and blunting oligodendrocyte death and loss of myelination as measured in the corpus callosum [674].

Neurodegenerative conditions themselves may provoke a neurosteroidogenic response as indicated in the case of dysmyelinating diseases. Progesterone levels increase in the brains of *shiverer* mice, which possess a deletion in the gene for myelin basic protein, and *jimpy* mice, whose phenotype is owed to a mutant proteolipid protein (PLP) and which models the X-linked conditions Pelizaeus-Merzbacher disease (PMD) and spastic paraplegia 2 (SPG2) [675].

Discrete changes in steroidogenic enzymes are also found in the substantia nigra and caudate nucleus of Parkinson's patients [676]. Treatment with estradiol, progesterone, or DHEA stifles the loss of dopamine in the methylphenyl tetrahydropyridine (MPTP) mouse model of Parkinsonism [677–680]. Estrogen's effect involves GPR30 and both ERs, with ERα being the most relevant [196, 681]. Mice lacking ERα are unresponsive to estradiol and more susceptible to MPTP [681]. Clinical studies with postmenopausal women indicate that estradiol reduces the risk of getting the disease, alleviates symptoms, and reduces the effective dose of levodopa [681]. These promising data correlate with the observation that Parkinson's disease is less prevalent in women. One study though reports that progesterone exacerbates the condition in hemiparkinsonian 6-hydroxydopamine (6-OHDA) model rats [682]. Thus, sex steroids are not universally beneficial.

Astonishingly, a single s.c. injection of allopregnanolone with β-cyclodextrin vehicle shortly after birth slows the onset and progression of Purkinje cell loss and cerebellar functional deficits in NPC mice, increasing lifespan nearly twofold [672]. Allopregnanolone increases myelination and reduces the expression of pro-inflammatory genes and infiltration of activated microglia into the cerebellum, possibly through PXR [683, 684]. While subsequent work concludes that the majority of the effect is due to β-cyclodextrin's ability to relieve intracellular cholesterol accumulation [685], the loss of neurosteroids with the dysfunction and death of highly steroidogenic Purkinje cells is considered to accelerate the course of the disease. The data thus indicate a therapeutic role for allopregnanolone.

It is unclear whether neurosteroids also promote neurodegenerative disease progression. One group speculates that increases in 3α-reduced neurosteroid synthesis and accumulation and the subsequent effects on the $GABA_A$ receptor serve such a function in encephalopathy and coma that accompany acute liver failure [686]. CNS loss of DHEA-S which antagonizes the receptor may allow the rise in 3α-reduced steroids predicted from experimental models to facilitate the onset of hepatic coma [686, 687].

13.2.1 Alzheimer's Disease (AD)

Allopregnanolone levels fall in the AD prefrontal cortex [688]. Changes in enzymes involved in its synthesis occur in this region and others [689]. At the same time, StAR is upregulated in hippocampal pyramidal neurons and astrocytes in AD compared to age-matched controls postmortem, suggesting attempts by neural cells to increase neurosteroid synthesis as part of an injury response [82]. Separate studies find no difference in StAR levels in the AD prefrontal cortex [689] nor for other steroidogenic genes in the AD hippocampus or cerebellum [690]. Other data show that DHEA levels in several regions of the AD brain are actually higher [688, 691], while pregnenolone sulfate and DHEA-S are reduced in the striatum and cerebellum [64]. The hypothalamus exhibits reduced DHEA-S [64] and increased pregnenolone [691].

Interestingly, regional differences in steroids observed in age-matched controls appear to vanish in the former survey. Thus, normal neurosteroidogenesis and steroid content are disrupted in the AD brain.

Allopregnanolone, estradiol, and androgens oppose the development and worsening of neuropathologic markers and behavioral and cognitive impairments in laboratory models [654, 692]. Allopregnanolone supports and restores neurogenesis and promotes neural cell survival in AD models [491, 693]. 10 mg/kg s.c. of the steroid fully rescues cognitive deficits and increases survival of newly formed cells in the hippocampus of triple transgenic AD mice at 6 and 9 months old, ages before and after β-amyloid plaque development [491]. At 12 months, the steroid is without effect.

Women have a higher incidence of AD [654]. This increased risk is theorized to be associated with the postmenopausal loss of sex steroids, such as estrogens [654]. In fact, polymorphisms in aromatase are linked to increased risk of sporadic AD in women and, in some studies, men [694]. Moreover, levels of aromatase and 17βHSD1 rise in the AD prefrontal cortex [689]. If StAR also rises in this structure as it does in the AD hippocampus, it suggests a neuroprotective change in neuroestradiol synthesis. Reductions in available androgen in aging males and consequent reduced CNS estradiol can also be linked to impaired cognition and increased AD risk [654, 695–697]. Both estrogenic and androgenic pathways are expected to contribute to neuroprotection.

Peripheral steroids can alter allopregnanolone synthesis and therefore perhaps bolster a neuroprotective response. Additionally, since in certain cases delineated in this review, systemic steroid administration mimics the action of neurosteroids, then perhaps peripheral steroids can substitute for them when neurosteroidogenesis is compromised. Thus, speculatively, as neuroprotective neurosteroids decline with neurodegeneration, the brain may rely on external steroids to fill in that void until that source too declines with age (assuming the changes are not concomitant or linked).

A final note is that levels of P450-7B1 also decline slightly in the CA1 and are almost halved in the dentate gyrus of the AD hippocampus [698]. Given 7α-hydroxypregnenolone's reported relevance to spatial memory [492], the loss of this enzyme may contribute to behavioral deficits in this disease.

13.2.2 HRT and AD

While aged patients with AD see short-term benefit of estradiol treatment, longer term treatment (12 vs. 2 months) may worsen the disease [497]. As predicted by the critical window model, the length of time following menopause before initiating HRT inversely correlates with estrogen's effectiveness in many aging and CNS injury models [654]. Furthermore, WHIMS reveals that estrogen plus MPA increases the risk of dementia and AD development in postmenopausal women [653]. This may reflect in part MPA opposition to neuroprotection by estrogen.

As mentioned before, continuous progesterone administration in HRT opposes the effectiveness of estradiol as a neuroprotectant. Studies in the triple transgenic model of AD replicate this finding but also reveal that cyclic progesterone treatment which mimics changes in cycling females boosts the effectiveness of estradiol and allows progesterone to assert its own neuroprotective effects [655]. In fact, HRT in OVX rats changes hippocampal PR levels in the CA1 in a manner dependent on the frequency of progesterone administration [699, 700]. Cyclic progesterone treatment with estradiol increases hippocampal PR content particularly in CA1 and especially using a cycle length (4 day) closer to the natural rodent cycle [699]. The use of continuous progesterone and/or estradiol reduces it. At the same time, PGRMC1, which promotes adult neuroprogenitor cell proliferation [171], rises in CA1, CA3, and the dentate gyrus with cyclic progesterone alone or in combination with estradiol [699, 700]. No other ER or progesterone target is known to change. Given the previously mentioned findings on cognition and cyclic estradiol treatment, one wonders whether cyclic estradiol would be of further benefit.

14 Neurosteroids and Convulsant Activity

14.1 Anticonvulsant Effects of Neurosteroids

As noted before, pro-convulsants can elicit increases in CNS steroid levels from neural or adrenal sources, likely as part of an injury and stress response. Androstanediol, allopregnanolone, and THDOC are potent anticonvulsants [103, 701–703]. The anticonvulsant properties of allopregnanolone and THDOC are mediated through the $GABA_A$ receptor. 3α-reduced steroids protect against many kinds of seizures and related neuronal damage in laboratory models [704]. Allopregnanolone protects against kainate and cocaine-induced seizures [609, 705] and kainate-induced limbic seizures and death [706]. Prior injection s.c. of pregnanolone or allopregnanolone counters pentylenetetrazol (PTZ)-induced seizures in mice [705]. Lower, ineffectual doses (below 3 mg/kg) of the steroids increase the efficacy of the anticonvulsant benzodiazepine diazepam (Valium) against PTZ. Allopregnanolone and THDOC also afford protection against NMDA-induced seizures and lethality and inhibit seizures provoked by $GABA_A$ receptor antagonists like picrotoxin and as noted, PTZ [705–708]. Pregnanolone also protects against NMDA, but at doses that alter motor activity [705].

As pointed out earlier, acute stress increases CNS allopregnanolone. The anxiolytic and anticonvulsant activities of the steroid explain the mechanism by which restraint stress opposes convulsant activity. The threshold dose for initiation of clonic seizures by a GABA channel blocker is increased when the animal undergoes restraint stress 30 min prior to treatment [709]. Application of either i.p. picrotoxin or s.c. finasteride prior to stress knock down the anticonvulsant effect of prior restraint stress [709]. Allopregnanolone s.c. or to some extent etifoxine i.p. administration prior to the stress overcomes inhibition by finasteride, consistent with the steroid being the cause of the anticonvulsant effect of stress. Restraint stress on the other hand has no effect on seizures induced by convulsants that act through kainate, NMDA, or glycine receptors.

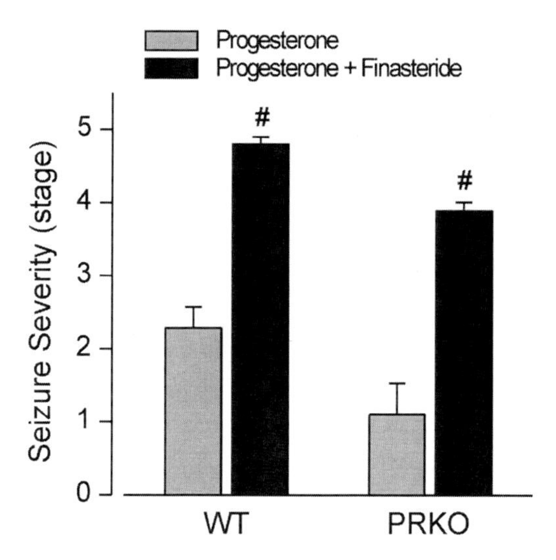

Fig. 16 *Progesterone protects against seizure severity in the amygdala-kindling model through metabolism to allopregnanolone.* Reductions in Racine seizure stage with kindling stimulation in fully kindled female mice given 75 mg/kg i.p. of progesterone 30 min prior are like wild-type (WT) mice, unaffected by the absence of a functional PR (PRKO). However, inhibition of 5α-reductase activity by i.p. finasteride pretreatment worsened seizure activity, indicating that progesterone must first be converted to allopregnanolone to exert a protective influence. Values represent mean ± SEM of 6–8 mice/group. #, $P < 0.05$ versus progesterone alone by Mann–Whitney U test (Adapted and reprinted with permission from [712])

Allopregnanolone and progesterone protect against secondarily generalized seizures in amygdala-kindling models of complex partial seizures [710]. Only at high i.p. doses (114 vs. 65 mg/kg) that induce ataxia is progesterone effective against focal seizures in half the animals in this study. On the other hand, s.c. DHP is effective against both components of amygdala-kindled seizures at a tenth of the dose and i.p. allopregnanolone only effective against secondarily generalized seizures at an ED_{50} of 15 mg/kg. The singular effectiveness of DHP here in males and in females in a related study [711] may rely on a separate membrane receptor. Data from the PR-knockout mouse indicate that the anticonvulsant effects of progesterone mainly depend upon its conversion to allopregnanolone [712] (Fig. 16). Inhibition of 5α-reductase activity by i.p. finasteride blocks protection by progesterone in the PTZ and amygdala-kindling models in wild-type and PR-null mice and correlates with the loss of allopregnanolone in the plasma (CNS levels were not examined) [712, 713]. On the other hand, progesterone uses a PR and allopregnanolone-independent mechanism to protect against maximal electroshock seizures [712].

14.1.1 Protection Against Epileptogenesis

Select steroids protect against pilocarpine-induced status epilepticus. Induction of seizure activity increases P450scc content in hippocampal CA3 neurons and especially astrocytes in the days following administration of the drug [35, 714]. Immunoreactivity for P450scc also arises in microglia and to a lesser extent, oligodendrocytes, consistent with the development of a remyelinating steroidogenic oligodendrocyte population. Expression of P450scc correlates with seizure duration in status epilepticus and delays in seizure onset. These findings imply that injury provokes neural synthesis of protective and restorative steroids.

Both allopregnanolone and THDOC protect against seizures and status epilepticus induced by kainate and pilocarpine [706], while s.c. finasteride reduces the time to seizure onset [35]. Similar to other injury models, the time at which allopregnanolone is supplied is important. Its ability to strongly potentiate $GABA_A$ inhibitory currents is transiently lost if given 24–48 h after the establishment of pilocarpine-induced status epilepticus [715]. The drop in allopregnanolone efficacy during this period may relate to the development of the disorder. The reduction in $GABA_A$ sensitivity is linked to changes in receptor subunit composition [716, 717].

Indeed, reductions in $GABA_A$ receptor-mediated tonic currents are linked to epileptogenesis [718] and loss of the δ subunit increases seizure susceptibility [313]. Reduced δ subunit levels in dentate gyrus neurons of *stargazer* mice decrease the sensitivity of tonic GABA currents to 3α-reduced steroids (THDOC), likely contributing to the animal's presentation of absence epilepsy [719].

A different example is the case of the Wistar Albino Glaxo from Rijswijk (WAG/Rij) rat. In this model of generalized absence epilepsy, rats present worsening seizure activity at 6 months. Central allopregnanolone and THDOC change independently from serum with age in male rats [720]. Thalamic levels of allopregnanolone are higher than Wistar controls at 2 months, but by 6, both steroids markedly decline in the thalamus, while serum and control thalamic levels remain unchanged and cerebrocortical levels rise. Thus, selective changes in thalamic neurosteroids correlate with the progression of absence epilepsy in this model.

In WAG/Rij rats, the decline in thalamic 3α,5α-reduced neurosteroids by 6 months of age coincides with a large increase in δ subunit levels along with elevated levels of α4 in select nuclei [720]. The combination enhances neuronal excitability, worsening absence seizures. Accordingly, allopregnanolone i.p. in 6-month-olds or microinjection of high pmol levels of allopregnanolone or its analog ganaxolone into some thalamic nuclei at 8–9 months increases seizure activity [721, 722]. Infusion into the perioral region of the primary somatosensory cortex of rats at the latter ages incidentally reduced spike-wave discharges [721]. Interestingly, reducing central allopregnanolone levels with long-term social isolation stress at 2 months reduces seizure activity at 6. Thus, the loss of atypically high allopregnanolone levels in the WAG/Rij CNS induces changes in $GABA_A$ receptor composition that support disease progression.

14.1.2 Seizures Related to Changes in Gonadal and Neural Steroids

The conclusion for the WAG/Rij rat model echoes the withdrawal model for PMS. Indeed, seizure threshold can be set by adrenal, gonadal, and neural steroids [704].

Increases in δ subunit levels in the mouse dentate gyrus related to the elevation in serum progesterone in late diestrus enhance resistance to seizures [311]. On the other hand, increased seizure frequency in patients with catamenial epilepsy is related to stages of the menstrual cycle at which there is little or a plunge in serum progesterone [104]. The drop raises seizure susceptibility, while progesterone supplementation is protective, likely preventing a withdrawal-related surge in α4 with δ.

This change is modeled in rats in a withdrawal paradigm that models PMS and postpartum syndrome. Withdrawal of long-term systemic progesterone administration decreases allopregnanolone levels while elevating α4 subunit expression, benzodiazepine-insensitive GABA-gated currents characteristic of increased δ subunit, and hippocampal CA1 neuronal excitability as mentioned before as well as seizure susceptibility [322]. The change in seizure susceptibility is suppressed by α4 antisense infusion into the right lateral ventricle. Other studies find similar effects with allopregnanolone withdrawal [704, 723].

Complementing these findings, clinical trials are testing ganaxolone supplementation as a specific treatment for catamenial epilepsy [104, 704]. Other studies target suppression of ovarian steroid output or use cyclic progestin supplementation [704]. All of these studies have met with varying levels of success to date.

The mechanism by which anticonvulsants in general work may involve their ability to increase allopregnanolone content as mentioned before, such as in the rat brain [351]. Following this lead, clinical research finds that ganaxolone shows promise alone or as an adjunctive drug in treating many types of seizure activity, like forms of refractory infantile spasms and pediatric epilepsy [705, 724, 725]. In a pilot study with hypogonadal epileptic men, testosterone supplementation reduces seizure frequency [726].

14.2 Pro-Convulsant Effects of Neurosteroids

Progesterone through PR increases seizure susceptibility as inferred, for example, from hippocampal and amygdalar kindling models in PR-knockout mice and with the use of RU486 [712, 727]. Loss of the receptor increases the anticonvulsant potency of progesterone presumably through its metabolism to allopregnanolone [712]. Allopregnanolone is not universally protective though. Like other $GABA_A$ receptor agonists, it exacerbates epileptogenesis induced by 4-aminopyridine [728].

14.2.1 Pregnenolone Sulfate

Sulfated neurosteroids are pro-convulsant. Injection of pregnenolone sulfate i.c.v. in the mouse or into the rat hippocampus induces seizures with ED_{50}s of 50 and 68 nmol, respectively, and, at high doses (4 µmol) in the rat hippocampus, status epilepticus [729, 730]. This likely results from the potentiation of NMDA receptor currents on target neurons and the antagonism of $GABA_A$ channels. Correspondingly, i.p. pregnenolone sulfate in the mouse increases the convulsant potency of NMDA, but not PTZ [731]. When infused i.c.v. at 50 nmol in the mouse, the potency of both PTZ and NMDA increases [729]. The effects of the steroid are blocked by dizocilpine, allopregnanolone, or a positive $GABA_A$ receptor modulator, clonazepam.

The effect of pregnenolone sulfate on convulsant activity changes with the dose and site of administration and, as noted, the seizure model. Infusion of 0.4–1 nmol into some thalamic nuclei and the perioral somatosensory cortex of 8–9-month-old WAG/Rij rats generally reduces absence seizures, while 0.2–0.4 nmol in these structures increases seizure activity [721]. Systemic administration in 6-month-olds elevates spike-wave discharges [722]. On the other hand, hippocampal coadministration of nicotine with a lower dose (5 ng or 12 pmol) of pregnenolone sulfate or 0.2 μg of allopregnanolone inhibits nicotine-induced audiogenic seizures in the rat, possibly through $GABA_A$ and neuronal nAchR [236]. While chronic treatment of mice with pregnenolone sulfate (or DHEA-S) increases the potency of PTZ, coadministration with progesterone is protective, possibly through its metabolism to allopregnanolone [732].

14.2.2 Estradiol, Testosterone, and HRT

Estradiol increases neuronal excitability and is generally pro-convulsant. Estradiol facilitates seizure activity induced by kainate in OVX rats [733] and induces seizure activity in CA1 pyramidal cells, likely through potentiation of EPSPs by NMDA receptors [734]. The pro-convulsant effects of estradiol are dose dependent and site specific [735]. For instance, in one of many studies that demonstrate the steroid's support of kindled seizures, estradiol facilitates the acquisition of kindled seizures in the dorsal hippocampus but not the ventral of OVX rats [736]. On the other hand, it protects against hippocampal damage associated with status epilepticus and ameliorates seizure activity in other models in OVX animals, often as a pretreatment [735, 737]. Another estrogen, estrone, inhibits kainate-induced seizures, toxicity, and lethality in male mice [738].

The impact of estradiol varies in a gender-specific manner and with gonadal status. Estradiol exposure increases neuronal excitability in hippocampal slices from male but not female rats [739]. Conversely, testosterone fails to increase excitability in slice cultures from males, but does so for diestrus females. Those from proestrus show reduced excitability.

Increased ovarian sources of estrogen versus progesterone contribute to catamenial-type seizures in epileptic perimenopausal women [735, 737]. As well, HRT can increase seizure frequency in menopausal women with epilepsy [737]. This outcome may arise not only from estrogen but the progestin component as well [659, 660]. Still, estrogen can exhibit anticonvulsant properties under certain conditions, such as is observed in various patients (summarized in [735]).

15 Roles in Tumors

The role of endogenous steroid production in tumor growth is an emerging area of cancer research. In the nervous system, brain tumors metabolize steroids [740], and as mentioned incidentally, many neural tumor cells are competent to produce steroids from cholesterol.

Several studies implicate neurosteroids in tumor progression. Brain tumors generally express StAR, but only those of glial origin [34]. Oligodendrogliomas and glial tumors express StAR at a higher density and level than in control white matter. Interestingly, the more malignant glioblastomas express higher levels of StAR than anaplastic astrocytomas. A follow-up study finds that oligodendrogliomas contain P450scc and StAR, but not 3βHSD [741]. Moreover, low-grade tumors express higher levels of StAR than high grade. Thus, neurosteroid synthesis may facilitate tumor growth. In the case of oligodendroglial tumors, the transition from low to high grade may reflect a sudden independence from the need for steroids to facilitate growth. The data are not unanimous however. An initial survey failed to identify P450scc in astrocytic tumors [92].

16 Neurosteroids and the Spinal Cord

The spinal cord expresses the enzymes necessary to synthesize steroid. Cervical, thoracic, lumbar, and sacral segments all contain levels of pregnenolone and progesterone far in excess of the serum in ADX/GDX and intact rats [742, 743]. Aside from the findings on pain transmission detailed later, the day-to-day function and regulation of steroids produced by spinal cord neurons unfortunately remains poorly understood. It is known though that spinal allopregnanolone production from progesterone is stimulated by glycine receptor activation [744].

Given the roles of neurosteroids in the developing brain, perhaps it is unsurprising that a protein dubbed VEMA (ventral midline antigen) which is implicated in the regulation of axon guidance during spinal cord development turns out to be PGRMC1 [745, 746]. Expression of PGRMC1 is primarily neuronal unlike PR and responds to progesterone with increased expression in male rats following complete spinal cord transection [168]. mPRs also localize in the spinal cord [658]. Future studies will determine the roles of these receptors in neuroprotection, function, and development.

16.1 Protective and Restorative Functions of Neurosteroids in Injury

Like the brain, spinal neurosteroids have neuroprotective and regenerative properties exerted through similar mechanisms with similar results, such as changes in neurotrophins and reduced inflammation. The rate of neurosteroid synthesis changes in response to injury. Transection of the spinal cord at the thoracic level in ADX/GDX or intact rats stimulates local pregnenolone and progesterone synthesis in the absence of changes in serum steroid [742]. Allopregnanolone is similarly elevated in the spinal cord of transected mice.

Basic research points to the clinical potential for neurosteroids in salvaging spinal cord function. Treatment of rats with spinal cord injuries (SCI) with progesterone improves outcomes and protects against white matter loss [747]. One astonishing study found that immediate administration of its precursor pregnenolone with anti-inflammatory and 3αHSD inhibitor indomethacin and lipopolysaccharide (LPS) following a controlled compressive injury to the spinal cord rescues motor abilities and traumatic tissue changes in a majority of affected rats [748]. More modest effects of the combined treatment are observed in a recent follow-up study [749]. Nevertheless, these results reveal a robust neuroprotective and restorative action by these steroids.

Other data support a neuroprotective and maturational role for androgens on motoneurons primarily through ARs [750]. The source of androgens in vivo is generally considered to be the serum. DHEA-S is also neuroprotective in spinal cord ischemia induced in rabbits, suggested to operate somehow through $GABA_A$ receptor potentiation [751]. DHEA-S and pregnenolone sulfate also inhibit glycine receptors in E17 rat spinal cord neurons, while the latter enhances NMDA receptor activity and inhibits other glutamate, glycine, and $GABA_A$ receptors in E7 chick spinal cord neurons [146, 148].

A different mechanism of neuroprotection is carried out by a synthetic pregnenolone analog that is resistant to metabolism to other steroid types. Administration of the drug s.c. within 24 h of injury improves the recovery of locomotor activity for three types of SCI in the rat [752]. While lesion size remains unchanged, the drug preserves dendrite arbor sizes in motoneurons and to a great extent, MAP2 levels through a direct interaction.

16.2 Neurosteroids in Amyotrophic Lateral Sclerosis (ALS) and Multiple Sclerosis (MS)

Both ALS and MS are diseases that attack the CNS. Studies on the use of neurosteroids in fighting these diseases have focused on spinal cord pathologies alongside those in the brain. Progesterone reduces aspects of spinal cord neurodegeneration and improves neuromuscular function and viability in *Wobbler* mice, a motoneuron disease model of idiopathic ALS [753, 754]. The mechanism of neuroprotection involves pro-myelinating effects reflected by elevated myelin basic protein levels, reduced astrogliosis, increased oligodendrocyte-precursor and oligodendrocyte counts, reduced local edema, and preservation of motoneurons in early and middle stages of the disease [747, 755, 756]. In a promising sign, a small ALS patient study positively correlates serum progesterone with slower disease progression and increased survival time [757]. Low progesterone associates with negative prognostic factors.

A role for steroids in muting microglial activation and CNS inflammation in MS is also under investigation [758]. Allopregnanolone and DHEA levels are reduced in white matter near lesions in the frontal lobe of MS patients and in the hindbrain of the rat experimental autoimmune encephalomyelitis (EAE) model of MS [759]. The loss of allopregnanolone may be accounted for by a sharp elevation in levels of microRNAs that target 3αHSD and corresponding reductions in common isoforms of the enzyme with MS and EAE. Pregnenolone levels are significantly increased in the EAE hindbrain, while 5α-reductase 1 mRNA is additionally reduced in MS. Thus, MS compromises allopregnanolone synthesis in the CNS at least in part through changes in key enzymes.

MS shows a female bias, especially postmenopause, while men have a worse prognosis – all implicating an influence of neuroactive sex steroids [760]. Indeed, allopregnanolone, progesterone, and estrogen reduce inflammation, motoneuron pathology, and other parameters in EAE rats [759–761]. Hence, the drop in neuroprotective allopregnanolone as well as DHEA in brain tissues may be a milestone in MS progression. Ongoing clinical trials include a focus on the use of progestin and estrogen therapies.

17 Neurosteroids and the PNS

The PNS collectively produces pregnenolone, progesterone, allopregnanolone, and estradiol. Satellite glia, Schwann cells, and the cell bodies of primary sensory neurons form one component, the DRG. Schwann cells do not normally generate steroids de novo, but neurons and satellite glia do [23, 27]. Neurosteroids then affect Schwann cell and neuronal function through nuclear and membrane receptors [762]. Developmentally, progesterone and allopregnanolone antagonize induction of $GABA_A$ receptor agonist muscimol of intracellular calcium transients in sensory neurons in the mouse DRG at E13, a stage when $GABA_A$ channel activity is excitatory [763]. Notably, these neurons also express PGRMC1 into adulthood.

It is not entirely clear what actions can be assigned to neurosteroids versus steroids from the periphery. Pelvic ganglia in the male rat express aromatase, and their maintenance is strongly influenced by castration in vivo and androgen and estradiol in vitro [764]. The biologically relevant source of estradiol is from local aromatization, but the leading source of substrate androgen is gonadal.

17.1 Schwann Cells and Myelination

Immature or myelinating Schwann cells in the DRG and the sciatic nerve are steroidogenic [30–32, 765]. The level of pregnenolone generated by such cells is sevenfold higher than in the circulation and remains high in nerves following castration and adrenalectomy [765–767]. But as with oligodendrocytes, Schwann cell maturation is associated with a loss of steroidogenic competence linked to declines in StAR and P450scc, but not 5α-reductase [27, 30].

In Schwann cells isolated from neonatal rat sciatic nerve and brachial plexus and grown separately from neurons, de novo production of P450scc and steroid correlates with dbcAMP-stimulated differentiation to a myelinating type [768]. Aminoglutethimide diminishes differentiation as gauged by levels of myelin protein 0 (P0/MPZ). Co-treatment with pregnenolone rescues it. Maturation after myelination is accompanied by a loss of steroidogenic potential [27], but conditions that invoke remyelination or reversion to a more immature cell type cause P450scc and 3βHSD expression to spike [30–32, 765]. Thus, neurosteroid production is an intrinsic part and marker of Schwann cell differentiation. Of note, StAR mRNA levels decline with robust cAMP stimulation in one Schwann cell culture system [30]. The authors of that study speculate that the effect of cAMP may be artificial since under conditions of injury, cAMP levels are very low.

Schwann cell neurosteroid production is part of the neuroprotective and regenerative response in the PNS [762]. Progesterone increases parameters of myelination in Schwann cell/DRG cocultures, such as number of myelinated segments and aggregate myelinated axon length [765]. Progesterone generated by Schwann cells increases the rate of myelination in part through increased production of myelin component proteins, like P0 or peripheral myelin protein-22 (PMP22) [31, 765, 769]. These effects are mediated by progesterone and DHP activation of PRs in the case of P0 and allopregnanolone likely through $GABA_A$ for PMP22 [762]. Progesterone and allopregnanolone also preserve the sciatic nerve against age-associated changes and loss of myelin fibers [770].

17.2 Protective and Regenerative Roles in Injury

Neurosteroids are protective with injury. Etifoxine improves sciatic nerve recovery from cryolesion, possibly through enhanced allopregnanolone synthesis [358, 771]. Inhibition of local progesterone formation presumably by Schwann cells or antagonism of PR following cryolesion hinders myelination of regenerating neurons [765]. Research further indicates the potential for progesterone to facilitate guided regeneration of facial nerves [772].

Estradiol production is upregulated in male rat DRG long term following sciatic nerve chronic constriction injury, a model of peripheral neuropathic pain [773]. Intrathecal letrozole causes apoptosis of satellite glia and with sciatic nerve injury, neuronal death. Thus, the estrogens generated are strongly protective. The same group found that DRG allopregnanolone and 3α-reductase are upregulated in the days following injury [774]. siRNA knockdown of the enzyme lowers nociceptive thresholds, pointing to a separate and important role for locally synthesized allopregnanolone.

17.3 Demyelinating Diseases and Diabetic Neuropathy

The *Trembler* mouse exhibits a demyelinating condition caused by PMP22 mutations that results in 16- and 6-fold lower concentrations of pregnenolone and progesterone, respectively, in the sciatic nerve [765]. Treatment with neurosteroids may thus represent a novel method to treat demyelinating conditions that affect P0 or PMP22 like Charcot-Marie-Tooth type 1B, hereditary neuropathy with liability to pressure palsies (HNPP), and Déjérine-Sottas disease. Interestingly, administration of a PR antagonist in a rat model of Charcot-Marie-Tooth type 1A disease decreases overexpression of PMP22, the cause of demyelination in this model, and improves clinical scores [775]. Progesterone worsens the disease phenotype.

In a streptozotocin (STZ) rat model of diabetic neuropathy, the sciatic nerve in males exhibits much lower levels of StAR, P450scc, and 5α-reductase mRNA and, correspondingly, pregnenolone, progesterone, DHP, testosterone, and androstanediol [776]. Weekly systemic treatment with a ligand for the transcription factor liver X receptor (LXR) rescues the expression of these genes, correcting the decline in steroids (except for testosterone) in the nerve but not the serum. Thermal sensitivity thresholds along with other measures of neuropathy are also restored, but not all myelin protein levels recover.

As well, supplementation with allopregnanolone, DHP, or progesterone improves aspects of peripheral nerve function and skin innervation including normalizing heat sensitivity thresholds in STZ rats, albeit by different receptor-mediated mechanisms [777]. Some parameters like improved myelin protein expression are only enhanced by the latter two steroids, suggesting the involvement of PR.

18 Adrenal Chromaffin Cells and PC12 Cells

Steroids influence adrenal chromaffin cells associated with the sympathetic nervous system and, as mentioned before, a developmental model for this cell type, PC12 cells. DHEA and DHEA-S differentially regulate proliferation of cultured bovine chromaffin cells from juvenile and adult cattle [778] and direct PC12 cells to differentiation to a neuroendocrine cell type, increasing basal dopamine release [779, 780]. Interestingly, PC12 cells lack NMDA and $GABA_A$ receptors, yet are responsive to pro-survival signaling by pregnenolone, allopregnanolone, DHEA, and DHEA-S. As noted previously, DHEA may exert its effects through binding of nerve growth factor receptors [267]. At the same time, activation of mARs in PC12 cells is pro-apoptotic [781]. Differentiation to a neuronal cell type dissipates mAR expression.

Practically, the measured effects of steroids on PC12 cells may not represent those of neurosteroids. In vivo, the steroids affecting chromaffin cells are likely to be of adrenal and serum not neural origin.

19 Integrated Actions of Neurosteroids in Nociception

Steroids regulate sensory pathways at many levels of the PNS and CNS. Gonadal steroids influence sensory processing and integration in the nervous system (for instance, see [782, 783]). Similar roles for endogenous steroids are currently being studied primarily in the context of nociception.

19.1 3α-Reduced Steroids and Thermal and Inflammatory Pain

Steroids can be analgesic at the level of the brain. I.c.v. allopregnanolone, 3αHP, and, to a lesser extent, progesterone increase thermal pain thresholds in male mice via a $GABA_A$ receptor-dependent mechanism [784, 785]. The effect of 3α-reduced neurosteroids is site specific. Intrahippocampal finasteride reduces allopregnanolone and tends to enhance analgesia, increasing latencies to tail flicks and paw licking in response to heat [341]. Administration of androstanediol into the POA also elicits anti-nociception through membrane receptors [786].

Substance P-containing cells under the skin are also steroidogenic, expressing P450scc [23]. In fact, P450scc expression follows along pain transmissive pathways to the brain, localizing in cell types in the DRG, the dorsal horn of the spinal cord, nociceptive supraspinal nuclei, and all the way up to the somatosensory cortex [26]. Production of allopregnanolone in peripheral nerves exerts analgesic effects through enhanced $GABA_A$ receptor activity and inhibition of T-type calcium channels [256, 787]. Afferents projecting onto sensory neurons in the spinal cord release substance P, which downregulates DHP and allopregnanolone formation from progesterone [788].

Spinal cord synthesis of allopregnanolone and THDOC selectively increases with hyperalgesia in response to thermal pain, leading to opposition of hyperalgesia through $GABA_A$ receptors at the level of lamina II interneurons and possibly supraspinal nuclei [789]. A separate study finds that micromolar intrathecal allopregnanolone and its 5β isomer decrease sensitivity to mechanical pressure in naive animals and allodynia likely through positive modulation of the $GABA_A$ receptor [790]. Of the two, only allopregnanolone opposes hyperalgesia generated by thermal heat in the λ-carrageenan model of inflammatory pain in male rats. The differential effect may owe to pregnanolone's opposition of glycine receptor activity.

In hind paw models of acute and persistent inflammatory pain, StAR, P450scc, and allopregnanolone levels selectively decline in the relevant portion of the spinal cord in mice [791]. Intraplantar injection of the antinocifensive PEA preserves or induces spinal levels of both proteins and allopregnanolone. The effect of PEA is partially stifled by i.p. aminoglutethimide or finasteride. Similarly, spinal StAR mRNA levels briefly spike in response to intraplantar LPS-induced peripheral inflammation in rats [20]. Intraspinal infusion of drugs that inhibit cyclooxygenase-2 (COX-2) or stabilize levels of an arachidonic acid metabolite epoxyeicosatrienoic acid and promote antihyperalgesia sustains the increase in StAR. This suggests that repression of COX-2 derepresses StAR and elevates neurosteroid synthesis as it does in testicular Leydig cells [20, 792].

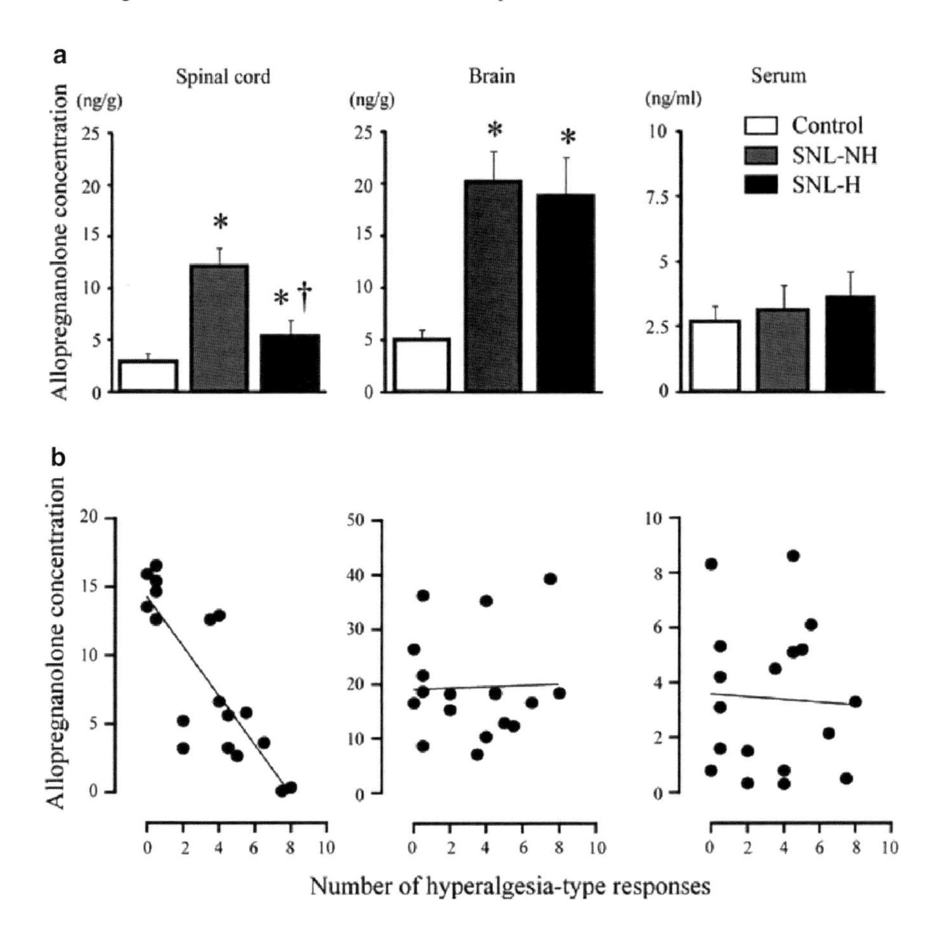

Fig. 17 *Spinal cord levels of allopregnanolone correlate with pain sensitivity with spinal nerve ligation (SNL).* (**a**) Allopregnanolone levels rise in the spinal cord and brain, but not the serum. The level was higher in the spinal cord for non-hyperalgesic animals, classified as those displaying 0–1 hyperalgesic-type responses to 10 separate needle pricks on the paw (SNL-NH). $*P < 0.05$ versus skin sham surgery control; †significantly lower ($P < 0.05$) versus SNL-NH group. (**b**) The level of allopregnanolone inversely correlates with the number of hyperalgesic responses out of 10 in individual animals. *Solid line* represents the result of simple linear regression with only the spinal cord graph exhibiting a significant relationship ($R^2 = 0.68$, slope $= -0.18$; $P < 0.0001$) (Adapted and reprinted with permission from [793])

19.2 Allopregnanolone and Neuropathic Pain

Lumbar spinal ligation, a model of neuropathic pain, increases allopregnanolone in the brain and the affected region of the spinal cord, but not the serum [793] (Fig. 17). Rats that present a lower hyperalgesic response in a subsequent test exhibit higher spinal levels of the steroid than those with a higher hyperalgesia score. No correlation was observed in the brain or the serum. Intrathecal infusion of allopregnanolone decreases the probability of a hyperalgesic response in rats that possess low levels of the steroid post-ligation, but not those with higher levels and little to no hyperalgesic response post-axotomy. Paroxetine s.c. is similarly antihyperalgesic in this model through selective increases in allopregnanolone in the spinal cord, but not the serum [794]. While the drug increases brain allopregnanolone in controls, it does not further increase brain content post-injury. Paroxetine's effect is not additive to allopregnanolone infusion and is countered by finasteride, all suggesting its main effects are through a feedback-regulated increase in allopregnanolone synthesis in the spinal cord.

19.3 DHEA, DHEA-S, and Pregnenolone Sulfate

Pregnenolone sulfate elicits nocifensive behavior and increases heat sensitivity in trigeminal and DRG sensory neurons via TRPM3 receptors [795]. Mice null for TRPM3 exhibit reduced populations of pregnenolone sulfate-responsive DRG and central trigeminal neurons and higher tolerance to noxious heat. Low levels of pregnenolone sulfate and DHEA-S induce nociceptive responses in mice through σ receptors and subsequent release of substance P at nerve endings [796]. Contrary to these results, a separate study found that intradermal injection of pregnenolone sulfate inhibits nocifensive responses by rats given capsaicin at the same site through TRPV1 receptors [797]. The reason for the discrepancy is unclear.

Intrathecal administration of DHEA-S potentiates NMDA receptor-dependent nociceptive behaviors through its positive regulation of σ receptors [798]. Effects on nociception with inflammatory pain via $P2X_2$ receptors are also suspected [244]. Progesterone blocks the action of both steroids [796].

DHEA opposes capsaicin-induced currents in DRG neurons [202]. At the same time, the steroid is pro-nociceptive within 15–30 min, but anti-nociceptive by 150 min in a rat model of neuropathic pain [799]. Its former effects involve NMDA receptor activation. Sciatic nerve ligature further lowers spinal levels of DHEA as well as the ability of spinal slices to generate the steroid from pregnenolone through reduced P450c17 expression. Blockade of P450c17 by intrathecal ketoconazole acutely provides analgesia. Since intrathecal testosterone is analgesic, increasing nociceptive thresholds in healthy male rats and those with sciatic nerve ligatures [799], it fits that the delayed analgesic effect of DHEA could reflect a pain-induced upregulation of enzymes in the relevant area of the spinal cord that convert it to testosterone.

19.4 Ovarian Steroids

Like testosterone, other sex steroids play a role in nociception. Pain thresholds are lower in women than men and change with the menstrual cycle [800, 801]. In a formalin model of pain, s.c. or i.c.v. ovarian steroids can be hyperalgesic unlike s.c. testosterone which has hypoalgesic effects independent of gender [802–804]. Enhanced sensitivity to capsaicin by female rats is lost with ovariectomy [805]. Estradiol potentiates the enhanced nocifensive response in OVX and male rats.

Mechanistically, estradiol potentiates TRPV1 currents in DRG neurons and capsaicin-induced nocifensive responses in male rats [202, 805]. Evidence from the ERβ-null mouse further supports the TRPV1 receptor as mediating estradiol's pro-nociceptive effects [806].

Estradiol can be anti-nociceptive as well, opposing intraplantar adjuvant-induced thermal and epinephrine-induced mechanical hyperalgesia [807, 808]. Estradiol application rapidly blocks L-type VGCCs via ERα and attenuates nociceptive ATP-induced intracellular calcium transients in rat DRG neurons [809, 810]. Progesterone is also anti-nociceptive at the level of the CNS through the modulation of κ and μ opioid receptors [811, 812]. Estradiol may indirectly affect the activity of these opioid receptors through membrane ERα. Progesterone also exerts nongenomic analgesic effects at the level of the VMH [786].

20 Mineralocorticoids and Glucocorticoids

Mineralocorticoids and glucocorticoids potently affect the CNS. High levels of glucocorticoids can damage neuronal function and are linked to adverse stress- and age-related changes in brain function [813, 814]. Indeed, glucocorticoid levels rise in the CSF with age [815]. Specific information on locally produced mineralocorticoids and glucocorticoids however is limited, partly because the effects of such steroids on the nervous system are commonly attributed to adrenal sources. Even CNS levels of THDOC in the rodent are clearly related to the adrenal [97]. That said, aldosterone and the main glucocorticoid in humans cortisol have very low penetrance through the blood–brain barrier [816]. Therefore, can all the effects of these steroids be attributed to the adrenal? Changes in CNS levels of aldosterone do follow changes in the serum, and removal of the adrenals results in a heavy loss of the steroid in the rat brain, but not a total loss [817, 818]. The remainder may reflect neural sources of steroid with physiologic relevance.

20.1 Glucocorticoids

While DOC is produced in the CNS, the enzyme responsible for its production, P450c21, is present at miniscule levels, including in rat astrocytes [3, 819–821], with mRNA levels estimated to be at least 10,000 times lower than in the adrenal [94, 821] (Fig. 1a). With the level of enzyme not seeming to match the level of 21-hydroxylase activity in the brain, other studies suggest that activity by isoforms of P4502D (rat 2D4 and human 2D6) make up the difference [822]. Such a possibility is not completely surprising since patients with nonfunctional P450c21 do retain some 21-hydroxylase activity [3]. A later estimate of transcript levels in the human up the range for P450c21 to between 1/100th and 1/1000th and for the enzyme integral for the final step in glucocorticoid synthesis P45011B1 to 1/100–1/10,000th that of the adrenal [19]. The latter proportion was also deduced for rat hippocampus.

Recent work along with previously noted data on P450scc and StAR verifies that male rat hippocampal neurons collectively possess the ability to fabricate corticosterone from cholesterol, expressing all the necessary enzymes including both P450c21 and P4502D4 [94]. Regional estimates do not clearly indicate that P45011B1 and P450c21 always co-express [19] and their levels vary between the hippocampus and hypothalamus [94]. Still, low nanomolar concentrations of corticosterone and DOC levels are detectable in ADX rats, levels sufficient to enhance spinogenesis [94]. Future research will illuminate what roles endogenous glucocorticoids play in the brain.

20.2 Aldosterone

The existence of neurally derived aldosterone remains controversial. Many different studies identify or dispute the existence of aldosterone synthase (P45011B2) in the brain across species [17, 19, 91, 690, 817, 823] (critiqued in [816]). One positive study found the enzyme to be discreetly distributed regionally, at levels around 3–4 orders of magnitude less than in the adrenal [19].

The function of neuroaldosterone may appropriately relate to hypertension. Hypothalamic aldosterone is elevated in the Dahl salt-sensitive rat model of hypertension [824]. Measured levels of P45011B2 are higher in select regions of the brains of the Dahl rat, while its levels in the adrenal are lower than in control rats [824]. Infusion of i.c.v. sodium increases hypothalamic but not hippocampal or serum aldosterone in rats [825]. The increase is blocked by co-infusion with the P45011B2 inhibitor FAD-286. Moreover, i.c.v. administration of FAD-286 selectively blocks angiotensin II-inspired increases in hypothalamic aldosterone without affecting serum levels of the steroid [826].

A pilot study finds that P450scc and 3βHSD are upregulated in the hypothalamus but not the adrenal of the Milan hypertensive rat [827]. The study's authors connect their increased expression to the formation of a ouabain-like factor, which later work disputes [828]. What the true steroid end product is remains unclear, but given that angiotensin II regulates this factor [829], candidates should include one regulated by this hormone in the adrenal zona glomerulosa and hypothalamus, aldosterone.

Known sites of action for mineralocorticoids in the brain are very limited. The salt-sensitive HSD2 neurons in the nucleus of the solitary tract are the best-described target, but they have access to circulating aldosterone due to an incomplete division between the blood and the brain in this region. Thus, any role for endogenous steroid here may be limited. Still, chronic infusion of trilostane or FAD-286 i.c.v. lowers blood pressure in Dahl rats on a high-salt diet, a diet that triggers hypertension in these animals [824, 830]. Thus, the increase in blood pressure with high salt may result from an increase in CSF sodium, which in turn stimulates central aldosterone synthesis [831]. Indeed, systemic s.c. infusion of the same low level of either drug does not affect blood pressure in the hypertensive Dahl rat, despite its potential diminution of adrenal aldosterone output. A P45011B1 inhibitor i.c.v. similarly blocks salt-induced hypertension in these animals [832]. As well, co-infusion of FAD-286 blocks the effects of i.c.v. sodium on hypertension in rats [825]. These results in total point to a separate role for neural aldosterone apart from the adrenal version in aspects of hypertension.

Mechanistically, aldosterone rapidly stimulates target cells through MRs at post-synaptic membranes, indicative of a role in neurotransmission [833, 834]. Corticosteroids may also utilize these receptors in preference to their cognate nuclear receptor [835]. The presence of 11βHSD type 2 sensitizes cells to aldosterone.

Interestingly, some effects of aldosterone such as stimulation of ERK1/2 phosphorylation may largely be mediated through GPR30 [836]. Recent findings in vascular smooth muscle cells show that the steroid activates GPR30 at picomolar levels, far lower than estrogen which acts in the nanomolar range [836]. Activation of GPR30 downregulates MR levels. Aldosterone further uses both MR and GPR30 to increase apoptosis.

21 Substance Abuse

Considerable research focuses on the influence of neurosteroids in substance addiction and abuse [11], a topic beyond the scope of this review. Steroids play an instrumental role in reward, tolerance, and withdrawal behavior. These effects are partly traced to their modulation of the $GABA_A$ receptor. The reward effects of estradiol are linked to the activation of ERs in the nucleus accumbens [837].

Steroids also mitigate impairments in CNS function induced by drugs, as mentioned previously. For instance, DHEA and pregnenolone sulfate among other steroids prevent cognitive deficits caused by ethanol [419]. Pregnenolone and its sulfate conjugate as well as high-dose DHEA oppose ethanol-induced anxiolysis [342, 345]. Select steroids also inhibit seizures induced by cocaine [705] and nicotine in response to an auditory stimulus [236].

Drugs themselves can regulate steroid levels in the CNS. Central levels of neurosteroids such as allopregnanolone rise with the intake of alcohol and Δ^9-tetrahydrocannabinol [11]. Chronic intermittent ethanol administration in rats reduces allopregnanolone along with its synthetic enzymes, 5α-reductase type I and 3αHSD [838]. This may be a cause of the increased anxiety and related behavioral changes observed in these animals.

Peripheral steroids such as those from the adrenal play a large role in drug effects and may contribute to or in some cases account for all the acute changes in CNS steroids. Roles for endogenous neurosteroids await future clarification.

22 Summary and Conclusions

Specific cell types in the nervous system generate steroids, not just from circulating precursors but de novo from cholesterol. Steroidogenic activity is scattered throughout the nervous system, with neurons accounting for most of it. With injury, however, comes a surge in neurosteroid synthesis by local glia, with estrogen being one major end product. Myelinating cells reacquire the ability to generate progesterone after substantially losing it during the maturation process. The production of the steroid seems integral for the protection and restoration of neuronal function.

The nature of steroidogenic stimulus is largely unknown. The existence of a novel membrane ERα/mGluR1a pathway in hypothalamic glia provides a unique mechanism by which peripheral estrogen can catalyze steroid production. At times, a single neural cell can generate many different steroid end products simultaneously. An example is the developing Purkinje cell which manufactures progesterone, allopregnanolone, and estrogen, steroids with overlapping functions [280]. How cells organize and manage the output of multiple steroids therefore becomes an intriguing question. Could steroid end products even vary between physical locations, such as axonal endings? An exciting array of possibilities exists.

An apparent contradiction is that while the effects of neurosteroids can be swift (seconds to minutes), their synthesis is comparatively slow, taking minutes to tens of minutes to generate significant concentrations of steroid. Unlike classic neurotransmitters, unconjugated steroids are not kept in secretory granules, ready to be dumped into pericellular spaces on demand. They freely diffuse across cell membranes, making storage difficult.

One can imagine two mechanisms to more quickly generate large amounts of a desired steroid. First, ongoing synthesis of one steroid could be diverted to another via, for instance, posttranslational modification of pertinent enzymes. Rapid changes (≤5 min) in estrogen production in the avian brain induced by glutamate may be so controlled through phosphorylation of aromatase [839]. Such regulation may apply in the human [839].

An alternative method is to increase general steroid production by modifying StAR. Studies in endocrine tissues find that StAR is not stored but, with or after its promotion of cholesterol transport to P450scc, is imported into the mitochondria, thereby terminating its activity (reviewed in [12]). Thus, continual steroid production requires continual production of StAR. To cut down on the time needed to upregulate steroidogenesis, a cell may simply increase the percentage of nascent StAR that is phosphorylated. This situation is observed in Leydig cells [68]. Activation of PKC increases levels of unphosphorylated StAR, which itself has at best a middling effect on steroid synthesis [67]. However, PKC potentiates a large steroidogenic response with co-activation of PKA, correlated to an increase in StAR phosphorylation [68].

The variety of steroids synthesized by neural cells is matched by the impressive number of membrane and nuclear receptor targets. Through synaptic and extrasynaptic receptors, neurosteroids modulate neuronal excitability and function through nongenomic pathways in the short term and genomic pathways longer term. Research has also only focused on a handful of the potential receptors with which neurosteroids can interact with in any given cell. Future efforts will define those that are actually bound and regulated by steroid in vivo and how they interact with other pathways.

The results of neurosteroids' actions on receptors are profound. The most heavily studied of the endogenous 3α-reduced steroids is allopregnanolone. This once overlooked steroid strongly potentiates inhibitory GABAergic currents, resulting in anxiolytic, sedative/hypnotic, anticonvulsant, analgesic, and neuroprotective effects. Endogenous levels of the steroid may preserve cognitive function, whereas exogenous supplementation is amnesic.

Falloffs or acute increases in its production alter $GABA_A$ receptor composition and localization through changes in α4 and δ subunits in target neurons, which can cause it to promote neuronal excitability. As a result, the steroid now has the opposite effect on seizures and anxiety. This becomes a recurring theme in behavioral studies on conditions that alter CNS allopregnanolone, such as puberty and PMS.

Gonad-typical steroid hormones like estrogen and progesterone utilize novel and classic receptors. An important discovery was that ERα can localize to the membrane to mediate nongenomic effects. Overall, estradiol is promnesic, pro-convulsant, neuroprotective, and generally hyperalgesic. Progesterone has mixed effects on memory, anxiety, and seizure activity, with the results of its stimulation of PR often in conflict with those from its metabolism to allopregnanolone. Both progesterone and allopregnanolone are also important in female sexual behavior and reproduction. One essential process, the LH surge, requires progesterone synthesized by hypothalamic glia.

Pregnenolone sulfate and DHEA variously affect σ1 and NMDA receptors, increasing neuronal excitability and opposing allopregnanolone. Sulfated steroids are generally anxiogenic, pro-convulsant, promnesic, and pro-nociceptive, that is, when they are not metabolized to allopregnanolone. They tend to oppose the actions of 3α-reduced steroids through the σ1 and NMDA receptors. One example is the conflicting effect that allopregnanolone and pregnenolone sulfate have on sleep. However, the true level and distribution of sulfated steroids and DHEA in the brain are an open question. Future research will also nail down roles for endogenous adrenal-type steroids and other more obscure ones like 7α-hydroxypregnenolone.

Drops in neurosteroid synthesis can follow the loss of gonadal steroids and neuroprotection along with behavioral and cognitive changes. Treatments like HRT may address the age-related shortfalls in CNS steroids and the problems entailed. The critical window hypothesis directs that HRT administration must be timely to salvage neuronal activity. Cyclic delivery of ovarian steroids to emulate the menstrual cycle may be key to maintaining the beneficial effects of HRT.

Collectively, studies on neurosteroids and their receptors support new approaches to treat conditions such as trauma, stroke, and seizures, and shed light on the mechanisms of older clinical therapies, like steroid replacement and SSRIs. They lend insight into injury and disease progression in the nervous system and in preventing, delaying, halting, and reversing damage. New perspectives on mood disorders and cognitive decline have also been uncovered. Questions linger over other day-to-day roles of neurosteroids. The widespread expression of neurosteroidogenic machinery implies broad involvement in many other nervous system functions, such as in sensory perception aside from pain, like auditory processing and olfaction.

Unfortunately, much of our knowledge about neurosteroids comes from experiments that fail to distinguish the source of the steroid, such as the use of i.p. administration. As well, many studies confusingly refer to peripheral neuroactive steroids and/or steroids derived from serum precursors as neurosteroids, in contravention of the original definition of the term [1, 3]. The use of inhibitors downstream of P450scc in vivo as well does not guarantee that the steroids examined wholly originate within the nervous system. A strict interpretation of the experimental data then would say we have many possible actions for neurosteroids, but few proven ones.

Further muddying the waters is the fact that the roles of both peripheral and local sources of steroid are deeply entangled. Systemic administration of steroids can duplicate or, in the case of injury, amplify the actions of neurosteroids. Similarly, peripheral steroids may normally support or lead such actions.

Coincidental changes in serum and neural levels of a given steroid add to the uncertainty. In the best example, estradiol in the hippocampus rises with estrus just like it does in the serum, but hippocampal estradiol largely originates from neural sources. Findings over the last decade indicate that neuroestradiol largely possesses the functions in cognition and neuroprotection in the hippocampus once attributed to gonadal estrogen. As research goes forward, more roles assumed to be mediated by peripheral steroids may be discovered to be handled by neurosteroids alone or complementarily. It makes sense given the positional edge local steroids have over peripheral.

Indeed, comparisons of steroid levels in the nervous system to the serum strongly indicate that at least for some structures like the hippocampus, local neurosteroids account for most or all of the activity attributable to steroids. The shortcoming of these assays is that they fail to define what constitutes a pharmacologic versus a physiologic level, since synaptic concentrations may be far greater than a regional assessment.

Given the apparent importance of neurosteroid synthesis in the CNS and PNS, one would expect confirmation in knockout models. However, studies on the loss of StAR and P450scc with patients and in animal models have yet to bear this out. Mutations in these genes cause congenital lipoid adrenal hyperplasia (CLAH) (reviewed in [12, 840]). In classic CLAH, catastrophic deficits in global steroid production result in primary adrenal insufficiency, male pseudohermaphroditism, and infertility. However, in both null animals and surveys of published patient data, no evidence exists of consistent structural abnormalities or behavioral deficits, aside from those deriving from compromised sexual differentiation and putative stress effects from adrenal insufficiency [12, 841]. These data tend to minimize the developmental and adult roles of all steroids, irrespective of origin, in the nervous system. On the other hand, the lack of a phenotype may represent a habituation to the absence of steroids.

That said, rigorous behavioral and cognitive testing is yet to be performed. Certainly, selective loss of one class of steroids does have an effect. Aromatase-knockout animals possess changes in anxiety and parental behaviors, addressable by estradiol supplementation (reviewed in [10]). As well, if a global loss of steroid increases the risk or accelerates the development of neurodegenerative diseases is unknown. If deficits are uncovered, the next question will then be whether the cause is the loss of neural or peripheral steroids. To this end, my laboratory developed a conditional knockout model for StAR. Early data suggest that cognition is unaffected with a lifelong targeted loss of StAR in the nervous system (unpublished observations, Whirledge SD, Smith AG, Lamb DJ, and King SR). H definitive experiments await.

Science's successes in the area of neurosteroids have impact beyond the ne system, shattering the notion that only endocrine production of steroids is physiologically relevant. Proof that steroids produced in small quantities exert important and rapid actions on local cells has led researchers to mine nonneural tissues for similar stories, like in the kidney [830].

Unfortunately, we are presently left with a fragmented picture of how neurosteroids fits into nervous system function. The next decade of research is set up to surprise and open new clinical horizons.

Acknowledgements The writing of this chapter was supported by NIH grant DK061548, the Laura W. Bush Institute for Women's Health and University Medical Center Women's Health Innovation Fund (WHIF), NIH Urology Training Grant Program T32 DK007763, Zorgniotti-Newman Prize of the International Society for Sexual Medicine, and a fellowship grant from the Lalor Foundation. The author would like to thank Dr. Peter J. Syapin for helpful discussions and his review of the manuscript.

End Note

*A recent study published at the time of printing reports that progesterone may extend its protection against damage from MCA primarily through PR rather than PGRMC1 [849].

References

1. Corpechot C, Robel P, Axelson M, Sjovall J, Baulieu EE (1981) Characterization and measurement of dehydroepiandrosterone sulfate in rat brain. Proc Natl Acad Sci USA 78:4704–4707
2. Corpechot C, Synguelakis M, Talha S, Axelson M, Sjovall J, Vihko R, Baulieu EE, Robel P (1983) Pregnenolone and its sulfate ester in the rat brain. Brain Res 270:119–125
3. Compagnone NA, Mellon SH (2000) Neurosteroids: biosynthesis and function of these novel neuromodulators. Front Neuroendocrinol 21:1–56
4. Cheney DL, Uzunov D, Costa E, Guidotti A (1995) Gas chromatographic-mass fragmentographic quantitation of 3 alpha-hydroxy-5 alpha-pregnan-20-one (allopregnanolone) and its precursors in blood and brain of adrenalectomized and castrated rats. J Neurosci 15:4641–4650
5. Robel P, Bourreau E, Corpechot C, Dang DC, Halberg F, Clarke C, Haug M, Schlegel ML, Synguelakis M, Vourch C (1987) Neuro-steroids: 3 beta-hydroxy-delta 5-derivatives in rat and monkey brain. J Steroid Biochem 27:649–655
6. Liere P, Pianos A, Eychenne B, Cambourg A, Liu S, Griffiths W, Schumacher M, Sjovall J, Baulieu EE (2004) Novel lipoidal derivatives of pregnenolone and dehydroepiandrosterone and absence of their sulfated counterparts in rodent brain. J Lipid Res 45:2287–2302
7. Liere P, Pianos A, Eychenne B, Cambourg A, Bodin K, Griffiths W, Schumacher M, Baulieu EE, Sjovall J (2009) Analysis of pregnenolone and dehydroepiandrosterone in rodent brain: cholesterol autoxidation is the key. J Lipid Res 50:2430–2444
8. Marx CE, Stevens RD, Shampine LJ, Uzunova V, Trost WT, Butterfield MI, Massing MW, Hamer RM, Morrow AL, Lieberman JA (2006) Neuroactive steroids are altered in schizophrenia and bipolar disorder: relevance to pathophysiology and therapeutics. Neuropsychopharmacology 31:1249–1263

9. Hojo Y, Higo S, Ishii H, Ooishi Y, Mukai H, Murakami G, Kominami T, Kimoto T, Honma S, Poirier D, Kawato S (2009) Comparison between hippocampus-synthesized and circulation-derived sex steroids in the hippocampus. Endocrinology 150:5106–5112
10. Do Rego JL, Seong JY, Burel D, Leprince J, Luu-The V, Tsutsui K, Tonon MC, Pelletier G, Vaudry H (2009) Neurosteroid biosynthesis: enzymatic pathways and neuroendocrine regulation by neurotransmitters and neuropeptides. Front Neuroendocrinol 30:259–301
11. Strous RD, Maayan R, Weizman A (2006) The relevance of neurosteroids to clinical psychiatry: from the laboratory to the bedside. Eur Neuropsychopharmacol 16:155–169
12. Bhangoo A, Anhalt H, Ten S, King SR (2006) Phenotypic variations in lipoid congenital adrenal hyperplasia. Pediatr Endocrinol Rev 3:258–271
13. Clark BJ, Wells J, King SR, Stocco DM (1994) The purification, cloning, and expression of a novel luteinizing hormone-induced mitochondrial protein in MA-10 mouse Leydig tumor cells. Characterization of the steroidogenic acute regulatory protein (StAR). J Biol Chem 269:28314–28322
14. Hu MC, Hsu NC, El Hadj NB, Pai CI, Chu HP, Wang CK, Chung BC (2002) Steroid deficiency syndromes in mice with targeted disruption of Cyp11a1. Mol Endocrinol 16:1943–1950
15. Hasegawa T, Zhao L, Caron KM, Majdic G, Suzuki T, Shizawa S, Sasano H, Parker KL (2000) Developmental roles of the steroidogenic acute regulatory protein (StAR) as revealed by StAR knockout mice. Mol Endocrinol 14:1462–1471
16. Furukawa A, Miyatake A, Ohnishi T, Ichikawa Y (1998) Steroidogenic acute regulatory protein (StAR) transcripts constitutively expressed in the adult rat central nervous system: colocalization of StAR, cytochrome P-450SCC (CYP XIA1), and 3beta-hydroxysteroid dehydrogenase in the rat brain. J Neurochem 71:2231–2238
17. Mellon SH, Deschepper CF (1993) Neurosteroid biosynthesis: genes for adrenal steroidogenic enzymes are expressed in the brain. Brain Res 629:283–292
18. Sanne JL, Krueger KE (1995) Expression of cytochrome P450 side-chain cleavage enzyme and 3 beta-hydroxysteroid dehydrogenase in the rat central nervous system: a study by polymerase chain reaction and in situ hybridization. J Neurochem 65:528–536
19. Yu L, Romero DG, Gomez-Sanchez CE, Gomez-Sanchez EP (2002) Steroidogenic enzyme gene expression in the human brain. Mol Cell Endocrinol 190:9–17
20. Inceoglu B, Jinks SL, Ulu A, Hegedus CM, Georgi K, Schmelzer KR, Wagner K, Jones PD, Morisseau C, Hammock BD (2008) Soluble epoxide hydrolase and epoxyeicosatrienoic acids modulate two distinct analgesic pathways. Proc Natl Acad Sci USA 105:18901–18906
21. King SR, Manna PR, Ishii T, Syapin PJ, Ginsberg SD, Wilson K, Walsh LP, Parker KL, Stocco DM, Smith RG, Lamb DJ (2002) An essential component in steroid synthesis, the steroidogenic acute regulatory protein, is expressed in discrete regions of the brain. J Neurosci 22:10613–10620
22. King SR, Ginsberg SD, Ishii T, Smith RG, Parker KL, Lamb DJ (2004) The steroidogenic acute regulatory protein is expressed in steroidogenic cells of the day-old brain. Endocrinology 145:4775–4780
23. Compagnone NA, Bulfone A, Rubenstein JL, Mellon SH (1995) Expression of the steroidogenic enzyme P450scc in the central and peripheral nervous systems during rodent embryogenesis. Endocrinology 136:2689–2696
24. Sierra A, Lavaque E, Perez-Martin M, Azcoitia I, Hales DB, Garcia-Segura LM (2003) Steroidogenic acute regulatory protein in the rat brain: cellular distribution, developmental regulation and overexpression after injury. Eur J Neurosci 18:1458–1467
25. Ukena K, Usui M, Kohchi C, Tsutsui K (1998) Cytochrome P450 side-chain cleavage enzyme in the cerebellar Purkinje neuron and its neonatal change in rats. Endocrinology 139:137–147
26. Patte-Mensah C, Kappes V, Freund-Mercier MJ, Tsutsui K, Mensah-Nyagan AG (2003) Cellular distribution and bioactivity of the key steroidogenic enzyme, cytochrome P450side chain cleavage, in sensory neural pathways. J Neurochem 86:1233–1246
27. Schaeffer V, Meyer L, Patte-Mensah C, Mensah-Nyagan AG (2010) Progress in dorsal root ganglion neurosteroidogenic activity: basic evidence and pathophysiological correlation. Prog Neurobiol 92:33–41

28. Gago N, Akwa Y, Sananes N, Guennoun R, Baulieu EE, El-Etr M, Schumacher M (2001) Progesterone and the oligodendroglial lineage: stage-dependent biosynthesis and metabolism. Glia 36:295–308
29. Schonemann MD, Muench MO, Tee MK, Miller WL, Mellon SH (2012) Expression of p450c17 in the human fetal nervous system. Endocrinology 153:2494–2505
30. Benmessahel Y, Troadec JD, Cadepond F, Guennoun R, Hales DB, Schumacher M, Groyer G (2004) Downregulation of steroidogenic acute regulatory protein (StAR) gene expression by cyclic AMP in cultured Schwann cells. Glia 45:213–228
31. Chan JR, Phillips LJ, Glaser M (1998) Glucocorticoids and progestins signal the initiation and enhance the rate of myelin formation. Proc Natl Acad Sci USA 95:10459–10464
32. Chan JR, Rodriguez-Waitkus PM, Ng BK, Liang P, Glaser M (2000) Progesterone synthesized by Schwann cells during myelin formation regulates neuronal gene expression. Mol Biol Cell 11:2283–2295
33. Saalmann YB, Kirkcaldie MT, Waldron S, Calford MB (2007) Cellular distribution of the GABAA receptor-modulating 3alpha-hydroxy, 5alpha-reduced pregnane steroids in the adult rat brain. J Neuroendocrinol 19:272–284
34. Kim HJ, Kim JE, Ha M, Kang SS, Kim JT, Park IS, Paek SH, Jung HW, Kim DG, Cho GJ, Choi WS (2003) Steroidogenic acute regulatory protein expression in the normal human brain and intracranial tumors. Brain Res 978:245–249
35. Biagini G, Longo D, Baldelli E, Zoli M, Rogawski MA, Bertazzoni G, Avoli M (2009) Neurosteroids and epileptogenesis in the pilocarpine model: evidence for a relationship between P450scc induction and length of the latent period. Epilepsia 50:53–58
36. Gottfried-Blackmore A, Sierra A, Jellinck PH, McEwen BS, Bulloch K (2008) Brain microglia express steroid-converting enzymes in the mouse. J Steroid Biochem Mol Biol 109:96–107
37. Guarneri P, Guarneri R, Cascio C, Pavasant P, Piccoli F, Papadopoulos V (1994) Neurosteroidogenesis in rat retinas. J Neurochem 63:86–96
38. Provost AC, Pequignot MO, Sainton KM, Gadin S, Salle S, Marchant D, Hales DB, Abitbol M (2003) Expression of SR-BI receptor and StAR protein in rat ocular tissues. C R Biol 326:841–851
39. Toyoshima K, Seta Y, Toyono T, Kataoka S (2007) Immunohistochemical identification of cells expressing steroidogenic enzymes cytochrome P450scc and P450 aromatase in taste buds of rat circumvallate papillae. Arch Histol Cytol 70:215–224
40. Lopez de Maturana R, Martin B, Millar RP, Brown P, Davidson L, Pawson AJ, Nicol MR, Mason JI, Barran P, Naor Z, Maudsley S (2007) GnRH-mediated DAN production regulates the transcription of the GnRH receptor in gonadotrope cells. Neuromolecular Med 9:230–248
41. Papadopoulos V (2004) In search of the function of the peripheral-type benzodiazepine receptor. Endocr Res 30:677–684
42. Decaudin D (2004) Peripheral benzodiazepine receptor and its clinical targeting. Anticancer Drugs 15:737–745
43. Papadopoulos V, Amri H, Boujrad N, Cascio C, Culty M, Garnier M, Hardwick M, Li H, Vidic B, Brown AS, Reversa JL, Bernassau JM, Drieu K (1997) Peripheral benzodiazepine receptor in cholesterol transport and steroidogenesis. Steroids 62:21–28
44. Papadopoulos V, Guarneri P, Kreuger KE, Guidotti A, Costa E (1992) Pregnenolone biosynthesis in C6-2B glioma cell mitochondria: regulation by a mitochondrial diazepam binding inhibitor receptor. Proc Natl Acad Sci USA 89:5113–5117
45. Chelli B, Falleni A, Salvetti F, Gremigni V, Lucacchini A, Martini C (2001) Peripheral-type benzodiazepine receptor ligands: mitochondrial permeability transition induction in rat cardiac tissue. Biochem Pharmacol 61:695–705
46. Hauet T, Yao ZX, Bose HS, Wall CT, Han Z, Li W, Hales DB, Miller WL, Culty M, Papadopoulos V (2005) Peripheral-type benzodiazepine receptor-mediated action of steroidogenic acute regulatory protein on cholesterol entry into Leydig cell mitochondria. Mol Endocrinol 19:540–554

47. King SR, Stocco DM (1996) ATP and a mitochondrial electrochemical gradient are required for functional activity of the steroidogenic acute regulatory (StAR) protein in isolated mitochondria. Endocr Res 22:505–514
48. King SR, Liu Z, Soh J, Eimerl S, Orly J, Stocco DM (1999) Effects of disruption of the mitochondrial electrochemical gradient on steroidogenesis and the Steroidogenic Acute Regulatory (StAR) protein. J Steroid Biochem Mol Biol 69:143–154
49. King SR, Walsh LP, Stocco DM (2000) Nigericin inhibits accumulation of the steroidogenic acute regulatory protein but not steroidogenesis. Mol Cell Endocrinol 166:147–153
50. King SR, Matassa AA, White EK, Walsh LP, Jo Y, Rao RM, Stocco DM, Reyland ME (2004) Oxysterols regulate expression of the steroidogenic acute regulatory protein. J Mol Endocrinol 32:507–517
51. Lutjohann D, Breuer O, Ahlborg G, Nennesmo I, Siden A, Diczfalusy U, Bjorkhem I (1996) Cholesterol homeostasis in human brain: evidence for an age-dependent flux of 24S-hydroxycholesterol from the brain into the circulation. Proc Natl Acad Sci USA 93:9799–9804
52. Hutson JC (2006) Physiologic interactions between macrophages and Leydig cells. Exp Biol Med (Maywood) 231:1–7
53. Cheney DL, Uzunov D, Guidotti A (1995) Pregnenolone sulfate antagonizes dizocilpine amnesia: role for allopregnanolone. Neuroreport 6:1697–1700
54. Romeo E, Cheney DL, Zivkovic I, Costa E, Guidotti A (1994) Mitochondrial diazepam-binding inhibitor receptor complex agonists antagonize dizocilpine amnesia: putative role for allopregnanolone. J Pharmacol Exp Ther 270:89–96
55. Ebner MJ, Corol DI, Havlikova H, Honour JW, Fry JP (2006) Identification of neuroactive steroids and their precursors and metabolites in adult male rat brain. Endocrinology 147:179–190
56. Kriz L, Bicikova M, Hill M, Hampl R (2005) Steroid sulfatase and sulfuryl transferase activity in monkey brain tissue. Steroids 70:960–969
57. Kriz L, Bicikova M, Mohapl M, Hill M, Cerny I, Hampl R (2008) Steroid sulfatase and sulfuryl transferase activities in human brain tumors. J Steroid Biochem Mol Biol 109:31–39
58. Shimada M, Yoshinari K, Tanabe E, Shimakawa E, Kobashi M, Nagata K, Yamazoe Y (2001) Identification of ST2A1 as a rat brain neurosteroid sulfotransferase mRNA. Brain Res 920:222–225
59. Kohjitani A, Fuda H, Hanyu O, Strott CA (2006) Cloning, characterization and tissue expression of rat SULT2B1a and SULT2B1b steroid/sterol sulfotransferase isoforms: divergence of the rat SULT2B1 gene structure from orthologous human and mouse genes. Gene 367:66–73
60. Kohjitani A, Fuda H, Hanyu O, Strott CA (2008) Regulation of SULT2B1a (pregnenolone sulfotransferase) expression in rat C6 glioma cells: relevance of AMPA receptor-mediated NO signaling. Neurosci Lett 430:75–80
61. Rajkowski KM, Robel P, Baulieu EE (1997) Hydroxysteroid sulfotransferase activity in the rat brain and liver as a function of age and sex. Steroids 62:427–436
62. Wang MD, Wahlstrom G, Backstrom T (1997) The regional brain distribution of the neurosteroids pregnenolone and pregnenolone sulfate following intravenous infusion. J Steroid Biochem Mol Biol 62:299–306
63. Schumacher M, Liere P, Akwa Y, Rajkowski K, Griffiths W, Bodin K, Sjovall J, Baulieu EE (2008) Pregnenolone sulfate in the brain: a controversial neurosteroid. Neurochem Int 52:522–540
64. Weill-Engerer S, David JP, Sazdovitch V, Liere P, Eychenne B, Pianos A, Schumacher M, Delacourte A, Baulieu EE, Akwa Y (2002) Neurosteroid quantification in human brain regions: comparison between Alzheimer's and nondemented patients. J Clin Endocrinol Metab 87:5138–5143
65. Mameli M, Carta M, Partridge LD, Valenzuela CF (2005) Neurosteroid-induced plasticity of immature synapses via retrograde modulation of presynaptic NMDA receptors. J Neurosci 25:2285–2294

66. Mellon SH, Griffin LD, Compagnone NA (2001) Biosynthesis and action of neurosteroids. Brain Res Brain Res Rev 37:3–12
67. Arakane F, King SR, Du Y, Kallen CB, Walsh LP, Watari H, Stocco DM, Strauss JF III (1997) Phosphorylation of steroidogenic acute regulatory protein (StAR) modulates its steroidogenic activity. J Biol Chem 272:32656–32662
68. Jo Y, King SR, Khan SA, Stocco DM (2005) Involvement of protein kinase C and cyclic adenosine 3′,5′-monophosphate-dependent kinase in steroidogenic acute regulatory protein expression and steroid biosynthesis in Leydig cells. Biol Reprod 73:244–255
69. Lavaque E, Mayen A, Azcoitia I, Tena-Sempere M, Garcia-Segura LM (2006) Sex differences, developmental changes, response to injury and cAMP regulation of the mRNA levels of steroidogenic acute regulatory protein, cytochrome p450scc, and aromatase in the olivocerebellar system. J Neurobiol 66:308–318
70. Manna PR, Chandrala SP, King SR, Jo Y, Counis R, Huhtaniemi IT, Stocco DM (2006) Molecular mechanisms of insulin-like growth factor-I mediated regulation of the steroidogenic acute regulatory protein in mouse Leydig cells. Mol Endocrinol 20:362–378
71. Roscetti G, Ambrosio C, Trabucchi M, Massotti M, Barbaccia ML (1994) Modulatory mechanisms of cyclic AMP-stimulated steroid content in rat brain cortex. Eur J Pharmacol 269:17–24
72. Karri S, Dertien JS, Stocco DM, Syapin PJ (2007) Steroidogenic acute regulatory protein expression and pregnenolone synthesis in rat astrocyte cultures. J Neuroendocrinol 19:860–869
73. Papadopoulos V, Guarneri P (1994) Regulation of C6 glioma cell steroidogenesis by adenosine 3′,5′-cyclic monophosphate. Glia 10:75–78
74. King SR, Stocco DM (2011) Steroidogenic acute regulatory (StAR) protein expression in the central nervous system. Front Neuroendocrine Sci 2:72
75. Morgan L, Jessen KR, Mirsky R (1991) The effects of cAMP on differentiation of cultured Schwann cells: progression from an early phenotype (04+) to a myelin phenotype (P0+, GFAP-, N-CAM-, NGF-receptor-) depends on growth inhibition. J Cell Biol 112:457–467
76. Raso GM, Esposito E, Vitiello S, Iacono A, Santoro A, D'Agostino G, Sasso O, Russo R, Piazza PV, Calignano A, Meli R (2011) Palmitoylethanolamide stimulation induces allopregnanolone synthesis in C6 Cells and primary astrocytes: involvement of peroxisome-proliferator activated receptor-alpha. J Neuroendocrinol 23:591–600
77. Sasso O, La Rana G, Vitiello S, Russo R, D'Agostino G, Iacono A, Russo E, Citraro R, Cuzzocrea S, Piazza PV, de Sarro G, Meli R, Calignano A (2010) Palmitoylethanolamide modulates pentobarbital-evoked hypnotic effect in mice: involvement of allopregnanolone biosynthesis. Eur Neuropsychopharmacol 20:195–206
78. Kushida A, Tamura H (2009) Retinoic acids induce neurosteroid biosynthesis in human glial GI-1 Cells via the induction of steroidogenic genes. J Biochem 146:917–923
79. Liu T, Wimalasena J, Bowen RL, Atwood CS (2007) Luteinizing hormone receptor mediates neuronal pregnenolone production via up-regulation of steroidogenic acute regulatory protein expression. J Neurochem 100:1329–1339
80. Rosati F, Sturli N, Cungi MC, Morello M, Villanelli F, Bartolucci G, Finocchi C, Peri A, Serio M, Danza G (2011) Gonadotropin-releasing hormone modulates cholesterol synthesis and steroidogenesis in SH-SY5Y cells. J Steroid Biochem Mol Biol 124:77–83
81. Lei ZM, Rao CV, Kornyei JL, Licht P, Hiatt ES (1993) Novel expression of human chorionic gonadotropin/luteinizing hormone receptor gene in brain. Endocrinology 132:2262–2270
82. Webber KM, Stocco DM, Casadesus G, Bowen RL, Atwood CS, Previll LA, Harris PL, Zhu X, Perry G, Smith MA (2006) Steroidogenic acute regulatory protein (StAR): evidence of gonadotropin-induced steroidogenesis in Alzheimer disease. Mol Neurodegener 1:14
83. Wilson AC, Salamat MS, Haasl RJ, Roche KM, Karande A, Meethal SV, Terasawa E, Bowen RL, Atwood CS (2006) Human neurons express type I GnRH receptor and respond to GnRH I by increasing luteinizing hormone expression. J Endocrinol 191:651–663
84. Barbaccia ML, Roscetti G, Bolacchi F, Concas A, Mostallino MC, Purdy RH, Biggio G (1996) Stress-induced increase in brain neuroactive steroids: antagonism by abecarnil. Pharmacol Biochem Behav 54:205–210

85. Barbaccia ML, Roscetti G, Trabucchi M, Purdy RH, Mostallino MC, Perra C, Concas A, Biggio G (1996) Isoniazid-induced inhibition of GABAergic transmission enhances neurosteroid content in the rat brain. Neuropharmacology 35:1299–1305
86. Do-Rego JL, Mensah-Nyagan GA, Beaujean D, Vaudry D, Sieghart W, Luu-The V, Pelletier G, Vaudry H (2000) Gamma-Aminobutyric acid, acting through gamma -aminobutyric acid type A receptors, inhibits the biosynthesis of neurosteroids in the frog hypothalamus. Proc Natl Acad Sci USA 97:13925–13930
87. Guarneri P, Russo D, Cascio C, de Leo G, Piccoli F, Guarneri R (1998) Induction of neurosteroid synthesis by NMDA receptors in isolated rat retina: a potential early event in excitotoxicity. Eur J Neurosci 10:1752–1763
88. Kimoto T, Tsurugizawa T, Ohta Y, Makino J, Tamura H, Hojo Y, Takata N, Kawato S (2001) Neurosteroid synthesis by cytochrome p450-containing systems localized in the rat brain hippocampal neurons: *N*-methyl-D-aspartate and calcium-dependent synthesis. Endocrinology 142:3578–3589
89. Guarneri P, Guarneri R, Cascio C, Piccoli F, Papadopoulos V (1995) Gamma-Aminobutyric acid type A/benzodiazepine receptors regulate rat retina neurosteroidogenesis. Brain Res 683:65–72
90. Do-Rego JL, Mensah-Nyagan AG, Feuilloley M, Ferrara P, Pelletier G, Vaudry H (1998) The endozepine triakontatetraneuropeptide diazepam-binding inhibitor [17–50] stimulates neurosteroid biosynthesis in the frog hypothalamus. Neuroscience 83:555–570
91. Pezzi V, Mathis JM, Rainey WE, Carr BR (2003) Profiling transcript levels for steroidogenic enzymes in fetal tissues. J Steroid Biochem Mol Biol 87:181–189
92. Inoue T, Akahira J, Suzuki T, Darnel AD, Kaneko C, Takahashi K, Hatori M, Shirane R, Kumabe T, Kurokawa Y, Satomi S, Sasano H (2002) Progesterone production and actions in the human central nervous system and neurogenic tumors. J Clin Endocrinol Metab 87:5325–5331
93. Agis-Balboa RC, Pinna G, Zhubi A, Maloku E, Veldic M, Costa E, Guidotti A (2006) Characterization of brain neurons that express enzymes mediating neurosteroid biosynthesis. Proc Natl Acad Sci USA 103:14602–14607
94. Higo S, Hojo Y, Ishii H, Komatsuzaki Y, Ooishi Y, Murakami G, Mukai H, Yamazaki T, Nakahara D, Barron A, Kimoto T, Kawato S (2011) Endogenous synthesis of corticosteroids in the hippocampus. PLoS One 6:e21631
95. Munetsuna E, Hojo Y, Hattori M, Ishii H, Kawato S, Ishida A, Kominami SA, Yamazaki T (2009) Retinoic acid stimulates 17beta-estradiol and testosterone synthesis in rat hippocampal slice cultures. Endocrinology 150:4260–4269
96. Mani S (2008) Progestin receptor subtypes in the brain: the known and the unknown. Endocrinology 149:2750–2756
97. Rupprecht R, Reul JM, Trapp T, van Steensel B, Wetzel C, Damm K, Zieglgansberger W, Holsboer F (1993) Progesterone receptor-mediated effects of neuroactive steroids. Neuron 11:523–530
98. Lamba V, Yasuda K, Lamba JK, Assem M, Davila J, Strom S, Schuetz EG (2004) PXR (NR1I2): splice variants in human tissues, including brain, and identification of neurosteroids and nicotine as PXR activators. Toxicol Appl Pharmacol 199:251–265
99. Rupprecht R, Holsboer F (1999) Neuroactive steroids: mechanisms of action and neuropsychopharmacological perspectives. Trends Neurosci 22:410–416
100. Fancsik A, Linn DM, Tasker JG (2000) Neurosteroid modulation of GABA IPSCs is phosphorylation dependent. J Neurosci 20:3067–3075
101. Belelli D, Lambert JJ (2005) Neurosteroids: endogenous regulators of the GABA(A) receptor. Nat Rev Neurosci 6:565–575
102. Gee KW, McCauley LD, Lan NC (1995) A putative receptor for neurosteroids on the GABAA receptor complex: the pharmacological properties and therapeutic potential of epalons. Crit Rev Neurobiol 9:207–227
103. Lambert JJ, Belelli D, Peden DR, Vardy AW, Peters JA (2003) Neurosteroid modulation of GABAA receptors. Prog Neurobiol 71:67–80

104. Reddy DS (2004) Role of neurosteroids in catamenial epilepsy. Epilepsy Res 62:99–118
105. Brot MD, Akwa Y, Purdy RH, Koob GF, Britton KT (1997) The anxiolytic-like effects of the neurosteroid allopregnanolone: interactions with GABA(A) receptors. Eur J Pharmacol 325:1–7
106. Weir CJ, Ling AT, Belelli D, Wildsmith JA, Peters JA, Lambert JJ (2004) The interaction of anaesthetic steroids with recombinant glycine and GABAA receptors. Br J Anaesth 92:704–711
107. Morrow AL, Suzdak PD, Paul SM (1987) Steroid hormone metabolites potentiate GABA receptor-mediated chloride ion flux with nanomolar potency. Eur J Pharmacol 142:483–485
108. Mohler H (2007) Molecular regulation of cognitive functions and developmental plasticity: impact of GABAA receptors. J Neurochem 102:1–12
109. Hosie AM, Wilkins ME, da Silva HM, Smart TG (2006) Endogenous neurosteroids regulate GABA(A)receptors through two discrete transmembrane sites. Nature 444:486–489
110. Hosie AM, Clarke L, da Silva H, Smart TG (2009) Conserved site for neurosteroid modulation of GABA A receptors. Neuropharmacology 56:149–154
111. Bracamontes JR, Steinbach JH (2009) Steroid interaction with a single potentiating site is sufficient to modulate GABA-A receptor function. Mol Pharmacol 75:973–981
112. Bracamontes J, McCollum M, Esch C, Li P, Ann J, Steinbach JH, Akk G (2011) Occupation of either site for the neurosteroid allopregnanolone potentiates the opening of the GABAA receptor induced from either transmitter binding site. Mol Pharmacol 80:79–86
113. Callachan H, Cottrell GA, Hather NY, Lambert JJ, Nooney JM, Peters JA (1987) Modulation of the GABAA receptor by progesterone metabolites. Proc R Soc Lond B Biol Sci 231:359–369
114. Mennerick S, He Y, Jiang X, Manion BD, Wang M, Shute A, Benz A, Evers AS, Covey DF, Zorumski CF (2004) Selective antagonism of 5alpha-reduced neurosteroid effects at GABA(A) receptors. Mol Pharmacol 65:1191–1197
115. Demirgoren S, Majewska MD, Spivak CE, London ED (1991) Receptor binding and electrophysiological effects of dehydroepiandrosterone sulfate, an antagonist of the GABAA receptor. Neuroscience 45:127–135
116. Majewska MD, Harrison NL, Schwartz RD, Barker JL, Paul SM (1986) Steroid hormone metabolites are barbiturate-like modulators of the GABA receptor. Science 232:1004–1007
117. Majewska MD, Schwartz RD (1987) Pregnenolone-sulfate: an endogenous antagonist of the gamma-aminobutyric acid receptor complex in brain? Brain Res 404:355–360
118. Majewska MD, Demirgoren S, Spivak CE, London ED (1990) The neurosteroid dehydroepiandrosterone sulfate is an allosteric antagonist of the GABAA receptor. Brain Res 526:143–146
119. Majewska MD, Mienville JM, Vicini S (1988) Neurosteroid pregnenolone sulfate antagonizes electrophysiological responses to GABA in neurons. Neurosci Lett 90:279–284
120. Calogero AE, Palumbo MA, Bosboom AM, Burrello N, Ferrara E, Palumbo G, Petraglia F, D'Agata R (1998) The neuroactive steroid allopregnanolone suppresses hypothalamic gonadotropin-releasing hormone release through a mechanism mediated by the gamma-aminobutyric acidA receptor. J Endocrinol 158:121–125
121. Akk G, Li P, Bracamontes J, Reichert DE, Covey DF, Steinbach JH (2008) Mutations of the GABA-A receptor alpha1 subunit M1 domain reveal unexpected complexity for modulation by neuroactive steroids. Mol Pharmacol 74:614–627
122. Park-Chung M, Malayev A, Purdy RH, Gibbs TT, Farb DH (1999) Sulfated and unsulfated steroids modulate gamma-aminobutyric acidA receptor function through distinct sites. Brain Res 830:72–87
123. Brussaard AB, Koksma JJ (2003) Conditional regulation of neurosteroid sensitivity of GABAA receptors. Ann N Y Acad Sci 1007:29–36
124. Belelli D, Casula A, Ling A, Lambert JJ (2002) The influence of subunit composition on the interaction of neurosteroids with GABA(A) receptors. Neuropharmacology 43:651–661
125. Rahman M, Lindblad C, Johansson IM, Backstrom T, Wang MD (2006) Neurosteroid modulation of recombinant rat alpha5beta2gamma2L and alpha1beta2gamma2L GABA(A) receptors in *Xenopus* oocyte. Eur J Pharmacol 547:37–44

126. Davies PA, Hanna MC, Hales TG, Kirkness EF (1997) Insensitivity to anaesthetic agents conferred by a class of GABA(A) receptor subunit. Nature 385:820–823
127. Caraiscos VB, Elliott EM, You-Ten KE, Cheng VY, Belelli D, Newell JG, Jackson MF, Lambert JJ, Rosahl TW, Wafford KA, MacDonald JF, Orser BA (2004) Tonic inhibition in mouse hippocampal CA1 pyramidal neurons is mediated by alpha5 subunit-containing gamma-aminobutyric acid type A receptors. Proc Natl Acad Sci USA 101:3662–3667
128. Hauser CA, Wetzel CH, Rupprecht R, Holsboer F (1996) Allopregnanolone acts as an inhibitory modulator on alpha1- and alpha6-containing GABA-A receptors. Biochem Biophys Res Commun 219:531–536
129. El-Etr M, Akwa Y, Fiddes RJ, Robel P, Baulieu EE (1995) A progesterone metabolite stimulates the release of gonadotropin-releasing hormone from GT1-1 hypothalamic neurons via the gamma-aminobutyric acid type A receptor. Proc Natl Acad Sci USA 92:3769–3773
130. Maitra R, Reynolds JN (1999) Subunit dependent modulation of GABAA receptor function by neuroactive steroids. Brain Res 819:75–82
131. Shingai R, Sutherland ML, Barnard EA (1991) Effects of subunit types of the cloned GABAA receptor on the response to a neurosteroid. Eur J Pharmacol 206:77–80
132. Shen H, Gong QH, Aoki C, Yuan M, Ruderman Y, Dattilo M, Williams K, Smith SS (2007) Reversal of neurosteroid effects at alpha4beta2delta GABAA receptors triggers anxiety at puberty. Nat Neurosci 10:469–477
133. Wohlfarth KM, Bianchi MT, Macdonald RL (2002) Enhanced neurosteroid potentiation of ternary GABA(A) receptors containing the delta subunit. J Neurosci 22:1541–1549
134. Stell BM, Brickley SG, Tang CY, Farrant M, Mody I (2003) Neuroactive steroids reduce neuronal excitability by selectively enhancing tonic inhibition mediated by delta subunit-containing GABAA receptors. Proc Natl Acad Sci USA 100:14439–14444
135. Peng Z, Hauer B, Mihalek RM, Homanics GE, Sieghart W, Olsen RW, Houser CR (2002) GABA(A) receptor changes in delta subunit-deficient mice: altered expression of alpha4 and gamma2 subunits in the forebrain. J Comp Neurol 446:179–197
136. Lovick TA, Griffiths JL, Dunn SM, Martin IL (2005) Changes in GABA(A) receptor subunit expression in the midbrain during the oestrous cycle in Wistar rats. Neuroscience 131:397–405
137. Poulter MO, Ohannesian L, Larmet Y, Feltz P (1997) Evidence that GABAA receptor subunit mRNA expression during development is regulated by GABAA receptor stimulation. J Neurochem 68:631–639
138. Zhang L, Chang YH, Feldman AN, Ma W, Lahjouji F, Barker JL, Hu Q, Maric D, Li BS, Li W, Rubinow DR (1999) The expression of GABA(A) receptor alpha2 subunit is upregulated by testosterone in rat cerebral cortex. Neurosci Lett 265:25–28
139. Shen H, Gong QH, Yuan M, Smith SS (2005) Short-term steroid treatment increases delta GABAA receptor subunit expression in rat CA1 hippocampus: pharmacological and behavioral effects. Neuropharmacology 49:573–586
140. Kuver A, Shen H, Smith SS (2012) Regulation of the surface expression of alpha4beta2delta GABA(A) receptors by high efficacy states. Brain Res 1463:1–20
141. Shingai R, Yanagi K, Fukushima T, Sakata K, Ogurusu T (1996) Functional expression of GABA rho 3 receptors in *Xenopus* oocytes. Neurosci Res 26:387–390
142. Morris KD, Moorefield CN, Amin J (1999) Differential modulation of the gamma-aminobutyric acid type C receptor by neuroactive steroids. Mol Pharmacol 56:752–759
143. Li W, Jin X, Covey DF, Steinbach JH (2007) Neuroactive steroids and human recombinant rho1 GABAC receptors. J Pharmacol Exp Ther 323:236–247
144. Wu FS, Gibbs TT, Farb DH (1990) Inverse modulation of gamma-aminobutyric acid- and glycine-induced currents by progesterone. Mol Pharmacol 37:597–602
145. Jiang P, Yang CX, Wang YT, Xu TL (2006) Mechanisms of modulation of pregnanolone on glycinergic response in cultured spinal dorsal horn neurons of rat. Neuroscience 141:2041–2050
146. Fodor L, Boros A, Dezso P, Maksay G (2006) Expression of heteromeric glycine receptor-channels in rat spinal cultures and inhibition by neuroactive steroids. Neurochem Int 49:577–583

147. Maksay G, Laube B, Betz H (2001) Subunit-specific modulation of glycine receptors by neurosteroids. Neuropharmacology 41:369–376
148. Wu FS, Gibbs TT, Farb DH (1991) Pregnenolone sulfate: a positive allosteric modulator at the *N*-methyl-D-aspartate receptor. Mol Pharmacol 40:333–336
149. Wu FS, Chen SC, Tsai JJ (1997) Competitive inhibition of the glycine-induced current by pregnenolone sulfate in cultured chick spinal cord neurons. Brain Res 750:318–320
150. Jiang P, Kong Y, Zhang XB, Wang W, Liu CF, Xu TL (2009) Glycine receptor in rat hippocampal and spinal cord neurons as a molecular target for rapid actions of 17-beta-estradiol. Mol Pain 5:2
151. Ziegler E, Bodusch M, Song Y, Jahn K, Wolfes H, Steinlechner S, Dengler R, Bufler J, Krampfl K (2009) Interaction of androsterone and progesterone with inhibitory ligand-gated ion channels: a patch clamp study. Naunyn Schmiedebergs Arch Pharmacol 380:277–291
152. Fishback JA, Robson MJ, Xu YT, Matsumoto RR (2010) Sigma receptors: potential targets for a new class of antidepressant drug. Pharmacol Ther 127:271–282
153. Martin WR, Eades CG, Thompson JA, Huppler RE, Gilbert PE (1976) The effects of morphine- and nalorphine- like drugs in the nondependent and morphine-dependent chronic spinal dog. J Pharmacol Exp Ther 197:517–532
154. Cahill MA (2007) Progesterone receptor membrane component 1: an integrative review. J Steroid Biochem Mol Biol 105:16–36
155. Su TP, London ED, Jaffe JH (1988) Steroid binding at sigma receptors suggests a link between endocrine, nervous, and immune systems. Science 240:219–221
156. Maurice T, Urani A, Phan VL, Romieu P (2001) The interaction between neuroactive steroids and the sigma1 receptor function: behavioral consequences and therapeutic opportunities. Brain Res Brain Res Rev 37:116–132
157. Maurice T, Roman FJ, Privat A (1996) Modulation by neurosteroids of the in vivo (+)-[3H] SKF-10,047 binding to sigma 1 receptors in the mouse forebrain. J Neurosci Res 46:734–743
158. Maurice T, Gregoire C, Espallergues J (2006) Neuro(active)steroids actions at the neuromodulatory sigma1 (sigma1) receptor: biochemical and physiological evidences, consequences in neuroprotection. Pharmacol Biochem Behav 84:581–597
159. Pal A, Chu UB, Ramachandran S, Grawoig D, Guo LW, Hajipour AR, Ruoho AE (2008) Juxtaposition of the steroid binding domain-like I and II regions constitutes a ligand binding site in the sigma-1 receptor. J Biol Chem 283:19646–19656
160. Schiess AR, Partridge LD (2005) Pregnenolone sulfate acts through a G-protein-coupled sigma1-like receptor to enhance short term facilitation in adult hippocampal neurons. Eur J Pharmacol 518:22–29
161. Bergeron R, de Montigny C, Debonnel G (1996) Potentiation of neuronal NMDA response induced by dehydroepiandrosterone and its suppression by progesterone: effects mediated via sigma receptors. J Neurosci 16:1193–1202
162. Debonnel G, Bergeron R, de Montigny C (1996) Potentiation by dehydroepiandrosterone of the neuronal response to *N*-methyl-D-aspartate in the CA3 region of the rat dorsal hippocampus: an effect mediated via sigma receptors. J Endocrinol 150:S33–S42
163. Monnet FP, Mahe V, Robel P, Baulieu EE (1995) Neurosteroids, via sigma receptors, modulate the [3H]norepinephrine release evoked by *N*-methyl-D-aspartate in the rat hippocampus. Proc Natl Acad Sci USA 92:3774–3778
164. Chen L, Sokabe M (2005) Presynaptic modulation of synaptic transmission by pregnenolone sulfate as studied by optical recordings. J Neurophysiol 94:4131–4144
165. Partridge LD, Valenzuela CF (2001) Neurosteroid-induced enhancement of glutamate transmission in rat hippocampal slices. Neurosci Lett 301:103–106
166. Phan VL, Su TP, Privat A, Maurice T (1999) Modulation of steroidal levels by adrenalectomy/castration and inhibition of neurosteroid synthesis enzymes affect sigma1 receptor-mediated behaviour in mice. Eur J Neurosci 11:2385–2396
167. Min L, Takemori H, Nonaka Y, Katoh Y, Doi J, Horike N, Osamu H, Raza FS, Vinson GP, Okamoto M (2004) Characterization of the adrenal-specific antigen IZA (inner zone antigen) and its role in the steroidogenesis. Mol Cell Endocrinol 215:143–148

168. Labombarda F, Gonzalez SL, Deniselle MC, Vinson GP, Schumacher M, de Nicola AF, Guennoun R (2003) Effects of injury and progesterone treatment on progesterone receptor and progesterone binding protein 25-Dx expression in the rat spinal cord. J Neurochem 87:902–913
169. Xu J, Zeng C, Chu W, Pan F, Rothfuss JM, Zhang F, Tu Z, Zhou D, Zeng D, Vangveravong S, Johnston F, Spitzer D, Chang KC, Hotchkiss RS, Hawkins WG, Wheeler KT, Mach RH (2011) Identification of the PGRMC1 protein complex as the putative sigma-2 receptor binding site. Nat Commun 2:380
170. Johannessen M, Fontanilla D, Mavlyutov T, Ruoho AE, Jackson MB (2011) Antagonist action of progesterone at sigma-receptors in the modulation of voltage-gated sodium channels. Am J Physiol Cell Physiol 300:C328–C337
171. Liu L, Wang J, Zhao L, Nilsen J, McClure K, Wong K, Brinton RD (2009) Progesterone increases rat neural progenitor cell cycle gene expression and proliferation via extracellularly regulated kinase and progesterone receptor membrane components 1 and 2. Endocrinology 150:3186–3196
172. Peluso JJ, Romak J, Liu X (2008) Progesterone receptor membrane component-1 (PGRMC1) is the mediator of progesterone's antiapoptotic action in spontaneously immortalized granulosa cells as revealed by PGRMC1 small interfering ribonucleic acid treatment and functional analysis of PGRMC1 mutations. Endocrinology 149:534–543
173. Guennoun R, Meffre D, Labombarda F, Gonzalez SL, Deniselle MC, Stein DG, de Nicola AF, Schumacher M (2008) The membrane-associated progesterone-binding protein 25-Dx: expression, cellular localization and up-regulation after brain and spinal cord injuries. Brain Res Rev 57:493–505
174. Rohe HJ, Ahmed IS, Twist KE, Craven RJ (2009) PGRMC1 (progesterone receptor membrane component 1): a targetable protein with multiple functions in steroid signaling, P450 activation and drug binding. Pharmacol Ther 121:14–19
175. Gerdes D, Wehling M, Leube B, Falkenstein E (1998) Cloning and tissue expression of two putative steroid membrane receptors. Biol Chem 379:907–911
176. Intlekofer KA, Petersen SL (2011) Distribution of mRNAs encoding classical progestin receptor, progesterone membrane components 1 and 2, serpine mRNA binding protein 1, and progestin and ADIPOQ receptor family members 7 and 8 in rat forebrain. Neuroscience 172:55–65
177. Spivak V, Lin A, Beebe P, Stoll L, Gentile L (2004) Identification of a neurosteroid binding site contained within the GluR2-S1S2 domain. Lipids 39:811–819
178. Park-Chung M, Wu FS, Farb DH (1994) 3 alpha-Hydroxy-5 beta-pregnan-20-one sulfate: a negative modulator of the NMDA-induced current in cultured neurons. Mol Pharmacol 46:146–150
179. Yaghoubi N, Malayev A, Russek SJ, Gibbs TT, Farb DH (1998) Neurosteroid modulation of recombinant ionotropic glutamate receptors. Brain Res 803:153–160
180. Wu FS, Yu HM, Tsai JJ (1998) Mechanism underlying potentiation by progesterone of the kainate-induced current in cultured neurons. Brain Res 779:354–358
181. Bowlby MR (1993) Pregnenolone sulfate potentiation of *N*-methyl-D-aspartate receptor channels in hippocampal neurons. Mol Pharmacol 43:813–819
182. Irwin RP, Maragakis NJ, Rogawski MA, Purdy RH, Farb DH, Paul SM (1992) Pregnenolone sulfate augments NMDA receptor mediated increases in intracellular Ca2+ in cultured rat hippocampal neurons. Neurosci Lett 141:30–34
183. Petrovic M, Sedlacek M, Horak M, Chodounska H, Vyklicky L Jr (2005) 20-oxo-5beta-pregnan-3alpha-yl sulfate is a use-dependent NMDA receptor inhibitor. J Neurosci 25:8439–8450
184. Horak M, Vlcek K, Petrovic M, Chodounska H, Vyklicky L Jr (2004) Molecular mechanism of pregnenolone sulfate action at NR1/NR2B receptors. J Neurosci 24:10318–10325
185. Horak M, Vlcek K, Chodounska H, Vyklicky L Jr (2006) Subtype-dependence of *N*-methyl-D-aspartate receptor modulation by pregnenolone sulfate. Neuroscience 137:93–102

186. Jang MK, Mierke DF, Russek SJ, Farb DH (2004) A steroid modulatory domain on NR2B controls *N*-methyl-D-aspartate receptor proton sensitivity. Proc Natl Acad Sci USA 101:8198–8203
187. Park-Chung M, Wu FS, Purdy RH, Malayev AA, Gibbs TT, Farb DH (1997) Distinct sites for inverse modulation of *N*-methyl-D-aspartate receptors by sulfated steroids. Mol Pharmacol 52:1113–1123
188. Malayev A, Gibbs TT, Farb DH (2002) Inhibition of the NMDA response by pregnenolone sulphate reveals subtype selective modulation of NMDA receptors by sulphated steroids. Br J Pharmacol 135:901–909
189. Johansson T, Frandberg PA, Nyberg F, Le Greves P (2008) Molecular mechanisms for nanomolar concentrations of neurosteroids at NR1/NR2B receptors. J Pharmacol Exp Ther 324:759–768
190. Elfverson M, Linde AM, Le Greves P, Zhou Q, Nyberg F, Johansson T (2008) Neurosteroids allosterically modulate the ion pore of the NMDA receptor consisting of NR1/NR2B but not NR1/NR2A. Biochem Biophys Res Commun 372:305–308
191. Johansson T, Le Greves P (2005) The effect of dehydroepiandrosterone sulfate and allopregnanolone sulfate on the binding of [(3)H]ifenprodil to the *N*-methyl-D-aspartate receptor in rat frontal cortex membrane. J Steroid Biochem Mol Biol 94:263–266
192. Compagnone NA, Mellon SH (1998) Dehydroepiandrosterone: a potential signalling molecule for neocortical organization during development. Proc Natl Acad Sci USA 95:4678–4683
193. Cyr M, Thibault C, Morissette M, Landry M, Di PT (2001) Estrogen-like activity of tamoxifen and raloxifene on NMDA receptor binding and expression of its subunits in rat brain. Neuropsychopharmacology 25:242–257
194. Smith CC, McMahon LL (2006) Estradiol-induced increase in the magnitude of long-term potentiation is prevented by blocking NR2B-containing receptors. J Neurosci 26:8517–8522
195. Weaver CE Jr, Park-Chung M, Gibbs TT, Farb DH (1997) 17beta-Estradiol protects against NMDA-induced excitotoxicity by direct inhibition of NMDA receptors. Brain Res 761:338–341
196. Morissette M, Le Saux M, D'Astous M, Jourdain S, Al Sweidi S, Morin N, Estrada-Camarena E, Mendez P, Garcia-Segura LM, Di Paolo T (2008) Contribution of estrogen receptors alpha and beta to the effects of estradiol in the brain. J Steroid Biochem Mol Biol 108:327–338
197. Le Saux M, Estrada-Camarena E, Di Paolo T (2006) Selective estrogen receptor-alpha but not -beta agonist treatment modulates brain alpha-amino-3-hydroxy-5-methyl-4-isoxazole-propionic acid receptors. J Neurosci Res 84:1076–1084
198. Morissette M, Le Saux M, Di Paolo T (2008) Effect of oestrogen receptor alpha and beta agonists on brain *N*-methyl-D-aspartate receptors. J Neuroendocrinol 20:1006–1014
199. Waters EM, Mitterling K, Spencer JL, Mazid S, McEwen BS, Milner TA (2009) Estrogen receptor alpha and beta specific agonists regulate expression of synaptic proteins in rat hippocampus. Brain Res 1290:1–11
200. Giuliani FA, Yunes R, Mohn CE, Laconi M, Rettori V, Cabrera R (2011) Allopregnanolone induces LHRH and glutamate release through NMDA receptor modulation. Endocrine 40:21–26
201. Wang C, Marx CE, Morrow AL, Wilson WA, Moore SD (2007) Neurosteroid modulation of GABAergic neurotransmission in the central amygdala: a role for NMDA receptors. Neurosci Lett 415:118–123
202. Chen SC, Chang TJ, Wu FS (2004) Competitive inhibition of the capsaicin receptor-mediated current by dehydroepiandrosterone in rat dorsal root ganglion neurons. J Pharmacol Exp Ther 311:529–536
203. Chen SC, Wu FS (2004) Mechanism underlying inhibition of the capsaicin receptor-mediated current by pregnenolone sulfate in rat dorsal root ganglion neurons. Brain Res 1027:196–200
204. Majeed Y, Amer MS, Agarwal AK, McKeown L, Porter KE, O'Regan DJ, Naylor J, Fishwick CW, Muraki K, Beech DJ (2011) Stereo-selective inhibition of transient receptor potential TRPC5 cation channels by neuroactive steroids. Br J Pharmacol 162:1509–1520

205. Majeed Y, Tumova S, Green BL, Seymour VA, Woods DM, Agarwal AK, Naylor J, Jiang S, Picton HM, Porter KE, O'Regan DJ, Muraki K, Fishwick CW, Beech DJ (2012) Pregnenolone sulphate-independent inhibition of TRPM3 channels by progesterone. Cell Calcium 51:1–11
206. Wagner TF, Loch S, Lambert S, Straub I, Mannebach S, Mathar I, Dufer M, Lis A, Flockerzi V, Philipp SE, Oberwinkler J (2008) Transient receptor potential M3 channels are ionotropic steroid receptors in pancreatic beta cells. Nat Cell Biol 10:1421–1430
207. Zamudio-Bulcock PA, Everett J, Harteneck C, Valenzuela CF (2011) Activation of steroid-sensitive TRPM3 channels potentiates glutamatergic transmission at cerebellar Purkinje neurons from developing rats. J Neurochem 119:474–485
208. Boulware MI, Weick JP, Becklund BR, Kuo SP, Groth RD, Mermelstein PG (2005) Estradiol activates group I and II metabotropic glutamate receptor signaling, leading to opposing influences on cAMP response element-binding protein. J Neurosci 25:5066–5078
209. Funakoshi T, Yanai A, Shinoda K, Kawano MM, Mizukami Y (2006) G protein-coupled receptor 30 is an estrogen receptor in the plasma membrane. Biochem Biophys Res Commun 346:904–910
210. Kelly MJ, Qiu J, Wagner EJ, Ronnekleiv OK (2002) Rapid effects of estrogen on G protein-coupled receptor activation of potassium channels in the central nervous system (CNS). J Steroid Biochem Mol Biol 83:187–193
211. Pedram A, Razandi M, Levin ER (2006) Nature of functional estrogen receptors at the plasma membrane. Mol Endocrinol 20:1996–2009
212. Prossnitz ER, Barton M (2011) The G-protein-coupled estrogen receptor GPER in health and disease. Nat Rev Endocrinol 7:715–726
213. Toran-Allerand CD, Guan X, MacLusky NJ, Horvath TL, Diano S, Singh M, Connolly ES Jr, Nethrapalli IS, Tinnikov AA (2002) ER-X: a novel, plasma membrane-associated, putative estrogen receptor that is regulated during development and after ischemic brain injury. J Neurosci 22:8391–8401
214. Kelly MJ, Loose MD, Ronnekleiv OK (1992) Estrogen suppresses mu-opioid- and GABAB-mediated hyperpolarization of hypothalamic arcuate neurons. J Neurosci 12:2745–2750
215. Levesque D, Di PT (1988) Rapid conversion of high into low striatal D2-dopamine receptor agonist binding states after an acute physiological dose of 17 beta-estradiol. Neurosci Lett 88:113–118
216. Qiu J, Bosch MA, Tobias SC, Grandy DK, Scanlan TS, Ronnekleiv OK, Kelly MJ (2003) Rapid signaling of estrogen in hypothalamic neurons involves a novel G-protein-coupled estrogen receptor that activates protein kinase C. J Neurosci 23:9529–9540
217. Schwarz S, Pohl P, Zhou GZ (1989) Steroid binding at sigma-"opioid" receptors. Science 246:1635–1638
218. Wagner EJ, Manzanares J, Moore KE, Lookingland KJ (1994) Neurochemical evidence that estrogen-induced suppression of kappa-opioid-receptor-mediated regulation of tuberoinfundibular dopaminergic neurons is prolactin-independent. Neuroendocrinology 59:197–201
219. Christensen HR, Zeng Q, Murawsky MK, Gregerson KA (2011) Estrogen regulation of the dopamine-activated GIRK channel in pituitary lactotrophs: implications for regulation of prolactin release during the estrous cycle. Am J Physiol Regul Integr Comp Physiol 301:R746–R756
220. Le Saux M, Morissette M, Di Paolo T (2006) ERbeta mediates the estradiol increase of D2 receptors in rat striatum and nucleus accumbens. Neuropharmacology 50:451–457
221. Schwarz S, Pohl P (1994) Steroids and opioid receptors. J Steroid Biochem Mol Biol 48:391–402
222. Roepke TA, Qiu J, Bosch MA, Ronnekleiv OK, Kelly MJ (2009) Cross-talk between membrane-initiated and nuclear-initiated oestrogen signalling in the hypothalamus. J Neuroendocrinol 21:263–270

223. Zhu Y, Bond J, Thomas P (2003) Identification, classification, and partial characterization of genes in humans and other vertebrates homologous to a fish membrane progestin receptor. Proc Natl Acad Sci USA 100:2237–2242
224. Thomas P, Pang Y, Dong J, Groenen P, Kelder J, de Vlieg J, Zhu Y, Tubbs C (2007) Steroid and G protein binding characteristics of the sea trout and human progestin membrane receptor alpha subtypes and their evolutionary origins. Endocrinology 148:705–718
225. Tang YT, Hu T, Arterburn M, Boyle B, Bright JM, Emtage PC, Funk WD (2005) PAQR proteins: a novel membrane receptor family defined by an ancient 7-transmembrane pass motif. J Mol Evol 61:372–380
226. Boonyaratanakornkit V, Bi Y, Rudd M, Edwards DP (2008) The role and mechanism of progesterone receptor activation of extra-nuclear signaling pathways in regulating gene transcription and cell cycle progression. Steroids 73:922–928
227. Sen A, Prizant H, Hammes SR (2011) Understanding extranuclear (nongenomic) androgen signaling: what a frog oocyte can tell us about human biology. Steroids 76:822–828
228. Tabori NE, Stewart LS, Znamensky V, Romeo RD, Alves SE, McEwen BS, Milner TA (2005) Ultrastructural evidence that androgen receptors are located at extranuclear sites in the rat hippocampal formation. Neuroscience 130:151–163
229. Lieberherr M, Grosse B (1994) Androgens increase intracellular calcium concentration and inositol 1,4,5-trisphosphate and diacylglycerol formation via a pertussis toxin-sensitive G-protein. J Biol Chem 269:7217–7223
230. Vicencio JM, Ibarra C, Estrada M, Chiong M, Soto D, Parra V, Az-Araya G, Jaimovich E, Lavandero S (2006) Testosterone induces an intracellular calcium increase by a nongenomic mechanism in cultured rat cardiac myocytes. Endocrinology 147:1386–1395
231. Pi M, Parrill AL, Quarles LD (2010) GPRC6A mediates the non-genomic effects of steroids. J Biol Chem 285:39953–39964
232. Klangkalya B, Chan A (1988) Inhibition of hypothalamic and pituitary muscarinic receptor binding by progesterone. Neuroendocrinology 47:294–302
233. Klangkalya B, Chan A (1988) Structure-activity relationships of steroid hormones on muscarinic receptor binding. J Steroid Biochem 29:111–118
234. Klangkalya B, Chan A (1988) The effects of ovarian hormones on beta-adrenergic and muscarinic receptors in rat heart. Life Sci 42:2307–2314
235. Horishita T, Minami K, Uezono Y, Shiraishi M, Ogata J, Okamoto T, Terada T, Sata T (2005) The effects of the neurosteroids: pregnenolone, progesterone and dehydroepiandrosterone on muscarinic receptor-induced responses in *Xenopus* oocytes expressing M1 and M3 receptors. Naunyn Schmiedebergs Arch Pharmacol 371:221–228
236. Martin-Garcia E, Pallares M (2005) The intrahippocampal administration of the neurosteroid allopregnanolone blocks the audiogenic seizures induced by nicotine. Brain Res 1062:144–150
237. Bullock AE, Clark AL, Grady SR, Robinson SF, Slobe BS, Marks MJ, Collins AC (1997) Neurosteroids modulate nicotinic receptor function in mouse striatal and thalamic synaptosomes. J Neurochem 68:2412–2423
238. Bertrand D, Valera S, Bertrand S, Ballivet M, Rungger D (1991) Steroids inhibit nicotinic acetylcholine receptors. Neuroreport 2:277–280
239. Valera S, Ballivet M, Bertrand D (1992) Progesterone modulates a neuronal nicotinic acetylcholine receptor. Proc Natl Acad Sci USA 89:9949–9953
240. Arias HR, Bhumireddy P, Bouzat C (2006) Molecular mechanisms and binding site locations for noncompetitive antagonists of nicotinic acetylcholine receptors. Int J Biochem Cell Biol 38:1254–1276
241. Ke L, Lukas RJ (1996) Effects of steroid exposure on ligand binding and functional activities of diverse nicotinic acetylcholine receptor subtypes. J Neurochem 67:1100–1112
242. Paradiso K, Zhang J, Steinbach JH (2001) The C terminus of the human nicotinic alpha4beta2 receptor forms a binding site required for potentiation by an estrogenic steroid. J Neurosci 21:6561–6568

243. Coddou C, Yan Z, Obsil T, Huidobro-Toro JP, Stojilkovic SS (2011) Activation and regulation of purinergic P2X receptor channels. Pharmacol Rev 63:641–683
244. De Roo M, Rodeau JL, Schlichter R (2003) Dehydroepiandrosterone potentiates native ionotropic ATP receptors containing the P2X2 subunit in rat sensory neurones. J Physiol 552:59–71
245. De Roo M, Boue-Grabot E, Schlichter R (2010) Selective potentiation of homomeric P2X2 ionotropic ATP receptors by a fast non-genomic action of progesterone. Neuropharmacology 58:569–577
246. Codocedo JF, Rodriguez FE, Huidobro-Toro JP (2009) Neurosteroids differentially modulate P2X ATP-gated channels through non-genomic interactions. J Neurochem 110:734–744
247. Wetzel CH, Hermann B, Behl C, Pestel E, Rammes G, Zieglgansberger W, Holsboer F, Rupprecht R (1998) Functional antagonism of gonadal steroids at the 5-hydroxytryptamine type 3 receptor. Mol Endocrinol 12:1441–1451
248. Zhang L, Sukhareva M, Barker JL, Maric D, Hao Y, Chang YH, Ma W, O'Shaughnessy T, Rubinow DR (2005) Direct binding of estradiol enhances Slack (sequence like a calcium-activated potassium channel) channels' activity. Neuroscience 131:275–282
249. Valverde MA, Rojas P, Amigo J, Cosmelli D, Orio P, Bahamonde MI, Mann GE, Vergara C, Latorre R (1999) Acute activation of Maxi-K channels (hSlo) by estradiol binding to the beta subunit. Science 285:1929–1931
250. Hou S, Heinemann SH, Hoshi T (2009) Modulation of BKCa channel gating by endogenous signaling molecules. Physiology (Bethesda) 24:26–35
251. Sarkar SN, Huang RQ, Logan SM, Yi KD, Dillon GH, Simpkins JW (2008) Estrogens directly potentiate neuronal L-type Ca2+ channels. Proc Natl Acad Sci USA 105:15148–15153
252. Druzin M, Malinina E, Grimsholm O, Johansson S (2011) Mechanism of estradiol-induced block of voltage-gated K+currents in rat medial preoptic neurons. PLoS One 6:e20213
253. Hige T, Fujiyoshi Y, Takahashi T (2006) Neurosteroid pregnenolone sulfate enhances glutamatergic synaptic transmission by facilitating presynaptic calcium currents at the calyx of Held of immature rats. Eur J Neurosci 24:1955–1966
254. Dong Y, Fu YM, Sun JL, Zhu YH, Sun FY, Zheng P (2005) Neurosteroid enhances glutamate release in rat prelimbic cortex via activation of alpha1-adrenergic and sigma1 receptors. Cell Mol Life Sci 62:1003–1014
255. Kobayashi T, Washiyama K, Ikeda K (2009) Pregnenolone sulfate potentiates the inwardly rectifying K channel Kir2.3. PLoS One 4:e6311
256. Pathirathna S, Brimelow BC, Jagodic MM, Krishnan K, Jiang X, Zorumski CF, Mennerick S, Covey DF, Todorovic SM, Jevtovic-Todorovic V (2005) New evidence that both T-type calcium channels and GABAA channels are responsible for the potent peripheral analgesic effects of 5alpha-reduced neuroactive steroids. Pain 114:429–443
257. Frye CA, Walf AA, Sumida K (2004) Progestins' actions in the VTA to facilitate lordosis involve dopamine-like type 1 and 2 receptors. Pharmacol Biochem Behav 78:405–418
258. Frye CA, Walf AA, Petralia SM (2006) In the ventral tegmental area, progestins have actions at D1 receptors for lordosis of hamsters and rats that involve GABA A receptors. Horm Behav 50:332–337
259. Dong LY, Cheng ZX, Fu YM, Wang ZM, Zhu YH, Sun JL, Dong Y, Zheng P (2007) Neurosteroid dehydroepiandrosterone sulfate enhances spontaneous glutamate release in rat prelimbic cortex through activation of dopamine D1 and sigma-1 receptor. Neuropharmacology 52:966–974
260. Dong Y, Zheng P (2011) Dehydroepiandrosterone sulfate: action and mechanism in the brain. J Neuroendocrinol 24:215–224
261. Feng XQ, Dong Y, Fu YM, Zhu YH, Sun JL, Wang Z, Sun FY, Zheng P (2004) Progesterone inhibition of dopamine-induced increase in frequency of spontaneous excitatory postsynaptic currents in rat prelimbic cortical neurons. Neuropharmacology 46:211–222
262. Bishop CV, Stormshak F (2006) Nongenomic action of progesterone inhibits oxytocin-induced phosphoinositide hydrolysis and prostaglandin F2alpha secretion in the ovine endometrium. Endocrinology 147:937–942

263. Bishop CV, Filtz T, Zhang Y, Slayden O, Stormshak F (2008) Progesterone suppresses an oxytocin-stimulated signal pathway in COS-7 cells transfected with the oxytocin receptor. Steroids 73:1367–1374
264. Grazzini E, Guillon G, Mouillac B, Zingg HH (1998) Inhibition of oxytocin receptor function by direct binding of progesterone. Nature 392:509–512
265. Bogacki M, Silvia WJ, Rekawiecki R, Kotwica J (2002) Direct inhibitory effect of progesterone on oxytocin-induced secretion of prostaglandin F(2alpha) from bovine endometrial tissue. Biol Reprod 67:184–188
266. Edwards DP (2005) Regulation of signal transduction pathways by estrogen and progesterone. Annu Rev Physiol 67:335–376
267. Lazaridis I, Charalampopoulos I, Alexaki VI, Avlonitis N, Pediaditakis I, Efstathopoulos P, Calogeropoulou T, Castanas E, Gravanis A (2011) Neurosteroid dehydroepiandrosterone interacts with nerve growth factor (NGF) receptors, preventing neuronal apoptosis. PLoS Biol 9:e1001051
268. Murakami K, Fellous A, Baulieu EE, Robel P (2000) Pregnenolone binds to microtubule-associated protein 2 and stimulates microtubule assembly. Proc Natl Acad Sci USA 97:3579–3584
269. Mizota K, Ueda H (2008) N-terminus of MAP2C as a neurosteroid-binding site. Neuroreport 19:1529–1533
270. Laurine E, Lafitte D, Gregoire C, Seree E, Loret E, Douillard S, Michel B, Briand C, Verdier JM (2003) Specific binding of dehydroepiandrosterone to the N terminus of the microtubule-associated protein MAP2. J Biol Chem 278:29979–29986
271. Kimoto T, Ishii H, Higo S, Hojo Y, Kawato S (2010) Semicomprehensive analysis of the postnatal age-related changes in the mRNA expression of sex steroidogenic enzymes and sex steroid receptors in the male rat hippocampus. Endocrinology 151:5795–5806
272. Zwain IH, Yen SS (1999) Neurosteroidogenesis in astrocytes, oligodendrocytes, and neurons of cerebral cortex of rat brain. Endocrinology 140:3843–3852
273. Pistritto G, Papacleovoulou G, Ragone G, Di Cesare S, Papaleo V, Mason JI, Barbaccia ML (2009) Differentiation-dependent progesterone synthesis and metabolism in NT2-N human neurons. Exp Neurol 217:302–311
274. Jung-Testas I, Hu ZY, Baulieu EE, Robel P (1989) Neurosteroids: biosynthesis of pregnenolone and progesterone in primary cultures of rat glial cells. Endocrinology 125:2083–2091
275. Le Goascogne C, Robel P, Gouezou M, Sananes N, Baulieu EE, Waterman M (1987) Neurosteroids: cytochrome P-450scc in rat brain. Science 237:1212–1215
276. Hu ZY, Bourreau E, Jung-Testas I, Robel P, Baulieu EE (1987) Neurosteroids: oligodendrocyte mitochondria convert cholesterol to pregnenolone. Proc Natl Acad Sci USA 84:8215–8219
277. Garcia CI, Paez PM, Soto EF, Pasquini JM (2007) Differential gene expression during development in two oligodendroglial cell lines overexpressing transferrin: A cDNA array analysis. Dev Neurosci 29:413–426
278. Sasahara K, Shikimi H, Haraguchi S, Sakamoto H, Honda S, Harada N, Tsutsui K (2007) Mode of action and functional significance of estrogen-inducing dendritic growth, spinogenesis, and synaptogenesis in the developing Purkinje cell. J Neurosci 27:7408–7417
279. Sakamoto H, Mezaki Y, Shikimi H, Ukena K, Tsutsui K (2003) Dendritic growth and spine formation in response to estrogen in the developing Purkinje cell. Endocrinology 144:4466–4477
280. Tsutsui K (2006) Biosynthesis and organizing action of neurosteroids in the developing Purkinje cell. Cerebellum 5:89–96
281. Sakamoto H, Ukena K, Tsutsui K (2001) Effects of progesterone synthesized de novo in the developing Purkinje cell on its dendritic growth and synaptogenesis. J Neurosci 21:6221–6232
282. Brinton RD (1994) The neurosteroid 3 alpha-hydroxy-5 alpha-pregnan-20-one induces cyto-architectural regression in cultured fetal hippocampal neurons. J Neurosci 14:2763–2774
283. Zhang L, Chang YH, Barker JL, Hu Q, Maric D, Li BS, Rubinow DR (2000) Testosterone and estrogen affect neuronal differentiation but not proliferation in early embryonic cortex of the rat: the possible roles of androgen and estrogen receptors. Neurosci Lett 281:57–60

284. de Lacalle S (2006) Estrogen effects on neuronal morphology. Endocrine 29:185–190
285. Spencer-Segal JL, Tsuda MC, Mattei L, Waters EM, Romeo RD, Milner TA, McEwen BS, Ogawa S (2012) Estradiol acts via estrogen receptors alpha and beta on pathways important for synaptic plasticity in the mouse hippocampal formation. Neuroscience 202:131–146
286. Zhang L, Li B, Zhao W, Chang YH, Ma W, Dragan M, Barker JL, Hu Q, Rubinow DR (2002) Sex-related differences in MAPKs activation in rat astrocytes: effects of estrogen on cell death. Brain Res Mol Brain Res 103:1–11
287. Guo J, Duckles SP, Weiss JH, Li X, Krause DN (2012) 17beta-Estradiol prevents cell death and mitochondrial dysfunction by an estrogen receptor-dependent mechanism in astrocytes after oxygen-glucose deprivation/reperfusion. Free Radic Biol Med 52:2151–2160
288. Galdo M, Gregonis J, Fiore CS, Compagnone NA (2012) Dehydroepiandrosterone biosynthesis, role, and mechanism of action in the developing neural tube. Front Endocrinol 3:1–15
289. Bologa L, Sharma J, Roberts E (1987) Dehydroepiandrosterone and its sulfated derivative reduce neuronal death and enhance astrocytic differentiation in brain cell cultures. J Neurosci Res 17:225–234
290. Suzuki M, Wright LS, Marwah P, Lardy HA, Svendsen CN (2004) Mitotic and neurogenic effects of dehydroepiandrosterone (DHEA) on human neural stem cell cultures derived from the fetal cortex. Proc Natl Acad Sci USA 101:3202–3207
291. Karishma KK, Herbert J (2002) Dehydroepiandrosterone (DHEA) stimulates neurogenesis in the hippocampus of the rat, promotes survival of newly formed neurons and prevents corticosterone-induced suppression. Eur J Neurosci 16:445–453
292. Fontaine-Lenoir V, Chambraud B, Fellous A, David S, Duchossoy Y, Baulieu EE, Robel P (2006) Microtubule-associated protein 2 (MAP2) is a neurosteroid receptor. Proc Natl Acad Sci USA 103:4711–4716
293. Mayo W, Lemaire V, Malaterre J, Rodriguez JJ, Cayre M, Stewart MG, Kharouby M, Rougon G, Le Moal M, Piazza PV, Abrous DN (2005) Pregnenolone sulfate enhances neurogenesis and PSA-NCAM in young and aged hippocampus. Neurobiol Aging 26:103–114
294. Wang JM, Johnston PB, Ball BG, Brinton RD (2005) The neurosteroid allopregnanolone promotes proliferation of rodent and human neural progenitor cells and regulates cell-cycle gene and protein expression. J Neurosci 25:4706–4718
295. Mtchedlishvili Z, Sun CS, Harrison MB, Kapur J (2003) Increased neurosteroid sensitivity of hippocampal GABAA receptors during postnatal development. Neuroscience 118:655–666
296. Grobin AC, Gizerian S, Lieberman JA, Morrow AL (2006) Perinatal allopregnanolone influences prefrontal cortex structure, connectivity and behavior in adult rats. Neuroscience 138:809–819
297. Gizerian SS, Moy SS, Lieberman JA, Grobin AC (2006) Neonatal neurosteroid administration results in development-specific alterations in prepulse inhibition and locomotor activity: neurosteroids alter prepulse inhibition and locomotor activity. Psychopharmacology (Berl) 186:334–342
298. Gizerian SS, Morrow AL, Lieberman JA, Grobin AC (2004) Neonatal neurosteroid administration alters parvalbumin expression and neuron number in medial dorsal thalamus of adult rats. Brain Res 1012:66–74
299. Mayo W, Le Moal M, Abrous DN (2001) Pregnenolone sulfate and aging of cognitive functions: behavioral, neurochemical, and morphological investigations. Horm Behav 40:215–217
300. Keller EA, Zamparini A, Borodinsky LN, Gravielle MC, Fiszman ML (2004) Role of allopregnanolone on cerebellar granule cells neurogenesis. Brain Res Dev Brain Res 153:13–17
301. Purdy RH, Morrow AL, Moore PH Jr, Paul SM (1991) Stress-induced elevations of gamma-aminobutyric acid type A receptor-active steroids in the rat brain. Proc Natl Acad Sci USA 88:4553–4557
302. Barbaccia ML, Concas A, Serra M, Biggio G (1998) Stress and neurosteroids in adult and aged rats. Exp Gerontol 33:697–712

303. Reddy DS, Kulkarni SK (1997) Differential anxiolytic effects of neurosteroids in the mirrored chamber behavior test in mice. Brain Res 752:61–71
304. Akwa Y, Purdy RH, Koob GF, Britton KT (1999) The amygdala mediates the anxiolytic-like effect of the neurosteroid allopregnanolone in rat. Behav Brain Res 106:119–125
305. Walf AA, Sumida K, Frye CA (2006) Inhibiting 5alpha-reductase in the amygdala attenuates antianxiety and antidepressive behavior of naturally receptive and hormone-primed ovariectomized rats. Psychopharmacology (Berl) 186:302–311
306. Verleye M, Akwa Y, Liere P, Ladurelle N, Pianos A, Eychenne B, Schumacher M, Gillardin JM (2005) The anxiolytic etifoxine activates the peripheral benzodiazepine receptor and increases the neurosteroid levels in rat brain. Pharmacol Biochem Behav 82:712–720
307. Bitran D, Dugan M, Renda P, Ellis R, Foley M (1999) Anxiolytic effects of the neuroactive steroid pregnanolone (3 alpha-OH-5 beta-pregnan-20-one) after microinjection in the dorsal hippocampus and lateral septum. Brain Res 850:217–224
308. Zimmerberg B, Rackow SH, George-Friedman KP (1999) Sex-dependent behavioral effects of the neurosteroid allopregnanolone (3alpha,5alpha-THP) in neonatal and adult rats after postnatal stress. Pharmacol Biochem Behav 64:717–724
309. Zimmerberg B, Brunelli SA, Fluty AJ, Frye CA (2005) Differences in affective behaviors and hippocampal allopregnanolone levels in adult rats of lines selectively bred for infantile vocalizations. Behav Brain Res 159:301–311
310. Zimmerberg B, Martinez AR, Skudder CM, Killien EY, Robinson SA, Brunelli SA (2010) Effects of gestational allopregnanolone administration in rats bred for high affective behavior. Physiol Behav 99:212–217
311. Maguire JL, Stell BM, Rafizadeh M, Mody I (2005) Ovarian cycle-linked changes in GABA(A) receptors mediating tonic inhibition alter seizure susceptibility and anxiety. Nat Neurosci 8:797–804
312. Laconi MR, Casteller G, Gargiulo PA, Bregonzio C, Cabrera RJ (2001) The anxiolytic effect of allopregnanolone is associated with gonadal hormonal status in female rats. Eur J Pharmacol 417:111–116
313. Mihalek RM, Banerjee PK, Korpi ER, Quinlan JJ, Firestone LL, Mi ZP, Lagenaur C, Tretter V, Sieghart W, Anagnostaras SG, Sage JR, Fanselow MS, Guidotti A, Spigelman I, Li Z, DeLorey TM, Olsen RW, Homanics GE (1999) Attenuated sensitivity to neuroactive steroids in gamma-aminobutyrate type A receptor delta subunit knockout mice. Proc Natl Acad Sci USA 96:12905–12910
314. Smith SS, Ruderman Y, Frye C, Homanics G, Yuan M (2006) Steroid withdrawal in the mouse results in anxiogenic effects of 3alpha,5beta-THP: a possible model of premenstrual dysphoric disorder. Psychopharmacology (Berl) 186:323–333
315. Martin-Garcia E, Darbra S, Pallares M (2008) Neonatal finasteride induces anxiogenic-like profile and deteriorates passive avoidance in adulthood after intrahippocampal neurosteroid administration. Neuroscience 154:1497–1505
316. Munetsuna E, Hattori M, Komatsu S, Sakimoto Y, Ishida A, Sakata S, Hojo Y, Kawato S, Yamazaki T (2009) Social isolation stimulates hippocampal estradiol synthesis. Biochem Biophys Res Commun 379:480–484
317. Agis-Balboa RC, Pinna G, Pibiri F, Kadriu B, Costa E, Guidotti A (2007) Down-regulation of neurosteroid biosynthesis in corticolimbic circuits mediates social isolation-induced behavior in mice. Proc Natl Acad Sci USA 104:18736–18741
318. Pibiri F, Nelson M, Guidotti A, Costa E, Pinna G (2008) Decreased corticolimbic allopregnanolone expression during social isolation enhances contextual fear: A model relevant for posttraumatic stress disorder. Proc Natl Acad Sci USA 105:5567–5572
319. Budziszewska B, Zajac A, Basta-Kaim A, Leskiewicz M, Steczkowska M, Lason W, Kacinski M (2010) Effects of neurosteroids on the human corticotropin-releasing hormone gene. Pharmacol Rep 62:1030–1040

320. Sarkar J, Wakefield S, MacKenzie G, Moss SJ, Maguire J (2011) Neurosteroidogenesis is required for the physiological response to stress: role of neurosteroid-sensitive GABAA receptors. J Neurosci 31:18198–18210
321. Miryala CS, Hassell J, Adams S, Hiegel C, Uzor N, Uphouse L (2011) Mechanisms responsible for progesterone's protection against lordosis-inhibiting effects of restraint II. Role of progesterone metabolites. Horm Behav 60:226–232
322. Smith SS, Gong QH, Hsu FC, Markowitz RS, ffrench-Mullen JM, Li X (1998) GABA(A) receptor alpha4 subunit suppression prevents withdrawal properties of an endogenous steroid. Nature 392:926–930
323. Aoki C, Sabaliauskas N, Chowdhury T, Min JY, Colacino AR, Laurino K, Barbarich-Marsteller NC (2012) Adolescent female rats exhibiting activity-based anorexia express elevated levels of GABA(A) receptor alpha4 and delta subunits at the plasma membrane of hippocampal CA1 spines. Synapse 66:391–407
324. Walf AA, Frye CA (2007) Administration of estrogen receptor beta-specific selective estrogen receptor modulators to the hippocampus decrease anxiety and depressive behavior of ovariectomized rats. Pharmacol Biochem Behav 86:407–414
325. Walf AA, Koonce C, Manley K, Frye CA (2009) Proestrous compared to diestrous wildtype, but not estrogen receptor beta knockout, mice have better performance in the spontaneous alternation and object recognition tasks and reduced anxiety-like behavior in the elevated plus and mirror maze. Behav Brain Res 196:254–260
326. Oyola MG, Portillo W, Reyna A, Foradori CD, Kudwa A, Hinds L, Handa RJ, Mani SK (2012) Anxiolytic effects and neuroanatomical targets of estrogen receptor-beta (ERbeta) activation by a selective ERbeta agonist in female mice. Endocrinology 153:837–846
327. Pluchino N, Luisi M, Lenzi E, Centofanti M, Begliuomini S, Freschi L, Ninni F, Genazzani AR (2006) Progesterone and progestins: Effects on brain, allopregnanolone and beta-endorphin. J Steroid Biochem Mol Biol 102:205–213
328. Genazzani AR, Stomati M, Bernardi F, Luisi S, Casarosa E, Puccetti S, Genazzani AD, Palumbo M, Luisi M (2004) Conjugated equine estrogens reverse the effects of aging on central and peripheral allopregnanolone and beta-endorphin levels in female rats. Fertil Steril 81:757–766
329. Lenzi E, Pluchino N, Begliuomini S, Casarosa E, Merlini S, Giannini A, Luisi M, Kumar N, Sitruk-Ware R, Genazzani AR (2009) Central modifications of allopregnanolone and beta-endorphin following subcutaneous administration of Nestorone. J Steroid Biochem Mol Biol 116:15–20
330. Pluchino N, Lenzi E, Merlini S, Giannini A, Cubeddu A, Casarosa E, Begliuomini S, Luisi M, Cela V, Genazzani AR (2009) Selective effect of chlormadinone acetate on brain allopregnanolone and opioids content. Contraception 80:53–62
331. Genazzani AR, Bernardi F, Stomati M, Monteleone P, Luisi S, Rubino S, Farzati A, Casarosa E, Luisi M, Petraglia F (2000) Effects of estradiol and raloxifene analog on brain, adrenal and serum allopregnanolone content in fertile and ovariectomized female rats. Neuroendocrinology 72:162–170
332. Stomati M, Bernardi F, Luisi S, Puccetti S, Casarosa E, Liut M, Quirici B, Pieri M, Genazzani AD, Luisi M, Genazzani AR (2002) Conjugated equine estrogens, estrone sulphate and estradiol valerate oral administration in ovariectomized rats: effects on central and peripheral allopregnanolone and beta-endorphin. Maturitas 43:195–206
333. Porcu P, Mostallino MC, Sogliano C, Santoru F, Berretti R, Concas A (2012) Long-term administration with levonorgestrel decreases allopregnanolone levels and alters GABA(A) receptor subunit expression and anxiety-like behavior. Pharmacol Biochem Behav 102:366–372
334. Frye CA, Sumida K, Dudek BC, Harney JP, Lydon JP, O'Malley BW, Pfaff DW, Rhodes ME (2006) Progesterone's effects to reduce anxiety behavior of aged mice do not require actions via intracellular progestin receptors. Psychopharmacology (Berl) 186:312–322
335. Reddy DS, Kulkarni SK (1996) Role of GABA-A and mitochondrial diazepam binding inhibitor receptors in the anti-stress activity of neurosteroids in mice. Psychopharmacology (Berl) 128:280–292

336. Reddy DS, Kulkarni SK (1997) Reversal of benzodiazepine inverse agonist FG 7142-induced anxiety syndrome by neurosteroids in mice. Methods Find Exp Clin Pharmacol 19:665–681
337. Reddy DS, O'Malley BW, Rogawski MA (2005) Anxiolytic activity of progesterone in progesterone receptor knockout mice. Neuropharmacology 48:14–24
338. Auger CJ, Forbes-Lorman RM (2008) Progestin receptor-mediated reduction of anxiety-like behavior in male rats. PLoS One 3:e3606
339. Pazol K, Wilson ME, Wallen K (2004) Medroxyprogesterone acetate antagonizes the effects of estrogen treatment on social and sexual behavior in female macaques. J Clin Endocrinol Metab 89:2998–3006
340. Pazol K, Northcutt KV, Patisaul HB, Wallen K, Wilson ME (2009) Progesterone and medroxyprogesterone acetate differentially regulate alpha4 subunit expression of GABA(A) receptors in the CA1 hippocampus of female rats. Physiol Behav 97:58–61
341. Rhodes ME, Frye CA (2001) Inhibiting progesterone metabolism in the hippocampus of rats in behavioral estrus decreases anxiolytic behaviors and enhances exploratory and antinociceptive behaviors. Cogn Affect Behav Neurosci 1:287–296
342. Melchior CL, Ritzmann RF (1994) Pregnenolone and pregnenolone sulfate, alone and with ethanol, in mice on the plus-maze. Pharmacol Biochem Behav 48:893–897
343. Meieran SE, Reus VI, Webster R, Shafton R, Wolkowitz OM (2004) Chronic pregnenolone effects in normal humans: attenuation of benzodiazepine-induced sedation. Psychoneuroendocrinology 29:486–500
344. Noda Y, Kamei H, Kamei Y, Nagai T, Nishida M, Nabeshima T (2000) Neurosteroids ameliorate conditioned fear stress: an association with sigma receptors. Neuropsychopharmacology 23:276–284
345. Melchior CL, Ritzmann RF (1994) Dehydroepiandrosterone is an anxiolytic in mice on the plus maze. Pharmacol Biochem Behav 47:437–441
346. Young J, Corpechot C, Haug M, Gobaille S, Baulieu EE, Robel P (1991) Suppressive effects of dehydroepiandrosterone and 3 beta-methyl-androst-5-en-17-one on attack towards lactating female intruders by castrated male mice. II. Brain neurosteroids. Biochem Biophys Res Commun 174:892–897
347. Avital A, Ram E, Maayan R, Weizman A, Richter-Levin G (2006) Effects of early-life stress on behavior and neurosteroid levels in the rat hypothalamus and entorhinal cortex. Brain Res Bull 68:419–424
348. Barbaccia ML, Roscetti G, Trabucchi M, Cuccheddu T, Concas A, Biggio G (1994) Neurosteroids in the brain of handling-habituated and naive rats: effect of CO2 inhalation. Eur J Pharmacol 261:317–320
349. Barbaccia ML, Roscetti G, Trabucchi M, Mostallino MC, Concas A, Purdy RH, Biggio G (1996) Time-dependent changes in rat brain neuroactive steroid concentrations and GABAA receptor function after acute stress. Neuroendocrinology 63:166–172
350. Dazzi L, Sanna A, Cagetti E, Concas A, Biggio G (1996) Inhibition by the neurosteroid allopregnanolone of basal and stress-induced acetylcholine release in the brain of freely moving rats. Brain Res 710:275–280
351. Jaworska-Feil L, Budziszewska B, Leskiewicz M, Lason W (2000) Effects of some centrally active drugs on the allopregnanolone synthesis in rat brain. Pol J Pharmacol 52:359–365
352. Griffin LD, Mellon SH (1999) Selective serotonin reuptake inhibitors directly alter activity of neurosteroidogenic enzymes. Proc Natl Acad Sci USA 96:13512–13517
353. Niwa T, Okada K, Hiroi T, Imaoka S, Narimatsu S, Funae Y (2008) Effect of psychotropic drugs on the 21-hydroxylation of neurosteroids, progesterone and allopregnanolone, catalyzed by rat CYP2D4 and human CYP2D6 in the brain. Biol Pharm Bull 31:348–351
354. Trauger JW, Jiang A, Stearns BA, LoGrasso PV (2002) Kinetics of allopregnanolone formation catalyzed by human 3 alpha-hydroxysteroid dehydrogenase type III (AKR1C2). Biochemistry 41:13451–13459
355. Pinna G, Costa E, Guidotti A (2006) Fluoxetine and norfluoxetine stereospecifically and selectively increase brain neurosteroid content at doses that are inactive on 5-HT reuptake. Psychopharmacology (Berl) 186:362–372

356. Nechmad A, Maayan R, Spivak B, Ramadan E, Poyurovsky M, Weizman A (2003) Brain neurosteroid changes after paroxetine administration in mice. Eur Neuropsychopharmacol 13:327–332
357. Uzunov DP, Cooper TB, Costa E, Guidotti A (1996) Fluoxetine-elicited changes in brain neurosteroid content measured by negative ion mass fragmentography. Proc Natl Acad Sci USA 93:12599–12604
358. Ugale RR, Sharma AN, Kokare DM, Hirani K, Subhedar NK, Chopde CT (2007) Neurosteroid allopregnanolone mediates anxiolytic effect of etifoxine in rats. Brain Res 1184:193–201
359. Marx CE, Duncan GE, Gilmore JH, Lieberman JA, Morrow AL (2000) Olanzapine increases allopregnanolone in the rat cerebral cortex. Biol Psychiatry 47:1000–1004
360. Marx CE, VanDoren MJ, Duncan GE, Lieberman JA, Morrow AL (2003) Olanzapine and clozapine increase the GABAergic neuroactive steroid allopregnanolone in rodents. Neuropsychopharmacology 28:1–13
361. Marx CE, Shampine LJ, Duncan GE, VanDoren MJ, Grobin AC, Massing MW, Madison RD, Bradford DW, Butterfield MI, Lieberman JA, Morrow AL (2006) Clozapine markedly elevates pregnenolone in rat hippocampus, cerebral cortex, and serum: candidate mechanism for superior efficacy? Pharmacol Biochem Behav 84:598–608
362. Barbaccia ML, Affricano D, Purdy RH, Maciocco E, Spiga F, Biggio G (2001) Clozapine, but not haloperidol, increases brain concentrations of neuroactive steroids in the rat. Neuropsychopharmacology 25:489–497
363. Pluchino N, Merlini S, Cubeddu A, Giannini A, Bucci F, Casarosa E, Cela V, Angioni S, Luisi M, Genazzani AR (2009) Brain-region responsiveness to DT56a (Femarelle) administration on allopregnanolone and opioid content in ovariectomized rats. Menopause 16:1037–1043
364. Genud R, Merenlender A, Gispan-Herman I, Maayan R, Weizman A, Yadid G (2009) DHEA lessens depressive-like behavior via GABA-ergic modulation of the mesolimbic system. Neuropsychopharmacology 34:577–584
365. Malkesman O, Shayit M, Genud R, Zangen A, Kinor N, Maayan R, Weizman A, Weller A, Yadid G (2007) Dehydroepiandrosterone in the nucleus accumbens is associated with early onset of depressive-behavior: a study in an animal model of childhood depression. Neuroscience 149:573–581
366. Morales AJ, Nolan JJ, Nelson JC, Yen SS (1994) Effects of replacement dose of dehydroepiandrosterone in men and women of advancing age. J Clin Endocrinol Metab 78:1360–1367
367. Wolkowitz OM, Reus VI, Roberts E, Manfredi F, Chan T, Ormiston S, Johnson R, Canick J, Brizendine L, Weingartner H (1995) Antidepressant and cognition-enhancing effects of DHEA in major depression. Ann N Y Acad Sci 774:337–339
368. Wolkowitz OM, Reus VI, Roberts E, Manfredi F, Chan T, Raum WJ, Ormiston S, Johnson R, Canick J, Brizendine L, Weingartner H (1997) Dehydroepiandrosterone (DHEA) treatment of depression. Biol Psychiatry 41:311–318
369. Reddy DS, Kaur G, Kulkarni SK (1998) Sigma (sigma1) receptor mediated anti-depressant-like effects of neurosteroids in the Porsolt forced swim test. Neuroreport 9:3069–3073
370. Urani A, Roman FJ, Phan VL, Su TP, Maurice T (2001) The antidepressant-like effect induced by sigma(1)-receptor agonists and neuroactive steroids in mice submitted to the forced swimming test. J Pharmacol Exp Ther 298:1269–1279
371. Dhir A, Kulkarni S (2008) Involvement of sigma (sigma1) receptors in modulating the anti-depressant effect of neurosteroids (dehydroepiandrosterone or pregnenolone) in mouse tail-suspension test. J Psychopharmacol 22:691–696
372. Bergeron R, de Montigny C, Debonnel G (1999) Pregnancy reduces brain sigma receptor function. Br J Pharmacol 127:1769–1776
373. Koss WA, Einat H, Schloesser RJ, Manji HK, Rubinow DR (2012) Estrogen effects on the forced swim test differ in two outbred rat strains. Physiol Behav 106:81–86
374. Maayan R, Abou-Kaud M, Strous RD, Kaplan B, Fisch B, Shinnar N, Weizman A (2004) The influence of parturition on the level and synthesis of sulfated and free neurosteroids in rats. Neuropsychobiology 49:17–23

375. Maayan R, Strous RD, Abou-Kaoud M, Weizman A (2005) The effect of 17beta estradiol withdrawal on the level of brain and peripheral neurosteroids in ovariectomized rats. Neurosci Lett 384:156–161
376. Sundstrom Poromaa I, Smith S, Gulinello M (2003) GABA receptors, progesterone and premenstrual dysphoric disorder. Arch Womens Ment Health 6:23–41
377. Uzunova V, Sheline Y, Davis JM, Rasmusson A, Uzunov DP, Costa E, Guidotti A (1998) Increase in the cerebrospinal fluid content of neurosteroids in patients with unipolar major depression who are receiving fluoxetine or fluvoxamine. Proc Natl Acad Sci USA 95:3239–3244
378. Kroboth PD, McAuley JW (1997) Progesterone: does it affect response to drug? Psychopharmacol Bull 33:297–301
379. Shirayama Y, Muneoka K, Fukumoto M, Tadokoro S, Fukami G, Hashimoto K, Iyo M (2011) Infusions of allopregnanolone into the hippocampus and amygdala, but not into the nucleus accumbens and medial prefrontal cortex, produce antidepressant effects on the learned helplessness rats. Hippocampus 21:1105–1113
380. Sundstrom PI, Smith S, Gulinello M (2003) GABA receptors, progesterone and premenstrual dysphoric disorder. Arch Womens Ment Health 6:23–41
381. Schmidt PJ, Nieman LK, Danaceau MA, Adams LF, Rubinow DR (1998) Differential behavioral effects of gonadal steroids in women with and in those without premenstrual syndrome. N Engl J Med 338:209–216
382. Genazzani AR, Palumbo MA, de Micheroux AA, Artini PG, Criscuolo M, Ficarra G, Guo AL, Benelli A, Bertolini A, Petraglia F, Purdy RH (1995) Evidence for a role for the neurosteroid allopregnanolone in the modulation of reproductive function in female rats. Eur J Endocrinol 133:375–380
383. Birzniece V, Turkmen S, Lindblad C, Zhu D, Johansson IM, Backstrom T, Wahlstrom G (2006) GABA(A) receptor changes in acute allopregnanolone tolerance. Eur J Pharmacol 535:125–134
384. Griffiths JL, Lovick TA (2005) GABAergic neurones in the rat periaqueductal grey matter express alpha4, beta1 and delta GABAA receptor subunits: plasticity of expression during the estrous cycle. Neuroscience 136:457–466
385. Gallo MA, Smith SS (1993) Progesterone withdrawal decreases latency to and increases duration of electrified prod burial: a possible rat model of PMS anxiety. Pharmacol Biochem Behav 46:897–904
386. Sundstrom I, Nyberg S, Backstrom T (1997) Patients with premenstrual syndrome have reduced sensitivity to midazolam compared to control subjects. Neuropsychopharmacology 17:370–381
387. Sundstrom I, Ashbrook D, Backstrom T (1997) Reduced benzodiazepine sensitivity in patients with premenstrual syndrome: a pilot study. Psychoneuroendocrinology 22:25–38
388. Shah NR, Jones JB, Aperi J, Shemtov R, Karne A, Borenstein J (2008) Selective serotonin reuptake inhibitors for premenstrual syndrome and premenstrual dysphoric disorder: a meta-analysis. Obstet Gynecol 111:1175–1182
389. Demetrio FN, Renno J Jr, Gianfaldoni A, Goncalves M, Halbe HW, Filho AH, Gorenstein C (2011) Effect of estrogen replacement therapy on symptoms of depression and anxiety in non-depressive menopausal women: a randomized double-blind, controlled study. Arch Womens Ment Health 14:479–486
390. Bain J (2010) Testosterone and the aging male: to treat or not to treat? Maturitas 66:16–22
391. Cubeddu A, Giannini A, Bucci F, Merlini S, Casarosa E, Pluchino N, Luisi S, Luisi M, Genazzani AR (2010) Paroxetine increases brain-derived neurotrophic factor in postmenopausal women. Menopause 17:338–343
392. Sarachana T, Xu M, Wu RC, Hu VW (2011) Sex hormones in autism: androgens and estrogens differentially and reciprocally regulate RORA, a novel candidate gene for autism. PLoS One 6:e17116
393. Biamonte F, Assenza G, Marino R, D'Amelio M, Panteri R, Caruso D, Scurati S, Yague JG, Garcia-Segura LM, Cesa R, Strata P, Melcangi RC, Keller F (2009) Interactions between

neuroactive steroids and reelin haploinsufficiency in Purkinje cell survival. Neurobiol Dis 36:103–115
394. Macri S, Biamonte F, Romano E, Marino R, Keller F, Laviola G (2010) Perseverative responding and neuroanatomical alterations in adult heterozygous reeler mice are mitigated by neonatal estrogen administration. Psychoneuroendocrinology 35:1374–1387
395. Baron-Cohen S, Lombardo MV, Auyeung B, Ashwin E, Chakrabarti B, Knickmeyer R (2011) Why are autism spectrum conditions more prevalent in males? PLoS Biol 9:e1001081
396. Pompili A, Arnone B, Gasbarri A (2012) Estrogens and memory in physiological and neuropathological conditions. Psychoneuroendocrinology 37:1379–1396
397. Marx CE, Bradford DW, Hamer RM, Naylor JC, Allen TB, Lieberman JA, Strauss JL, Kilts JD (2011) Pregnenolone as a novel therapeutic candidate in schizophrenia: emerging preclinical and clinical evidence. Neuroscience 191:78–90
398. Darbra S, Modol L, Pallares M (2012) Allopregnanolone infused into the dorsal (CA1) hippocampus increases prepulse inhibition of startle response in Wistar rats. Psychoneuroendocrinology 37:581–585
399. Umathe SN, Vaghasiya JM, Jain NS, Dixit PV (2009) Neurosteroids modulate compulsive and persistent behavior in rodents: implications for obsessive-compulsive disorder. Prog Neuropsychopharmacol Biol Psychiatry 33:1161–1166
400. Albelda N, Joel D (2012) Animal models of obsessive-compulsive disorder: exploring pharmacology and neural substrates. Neurosci Biobehav Rev 36:47–63
401. Hill RA, McInnes KJ, Gong EC, Jones ME, Simpson ER, Boon WC (2007) Estrogen deficient male mice develop compulsive behavior. Biol Psychiatry 61:359–366
402. Roy-Byrne PP, Cowley DS, Greenblatt DJ, Shader RI, Hommer D (1990) Reduced benzodiazepine sensitivity in panic disorder. Arch Gen Psychiatry 47:534–538
403. Roy-Byrne P, Wingerson DK, Radant A, Greenblatt DJ, Cowley DS (1996) Reduced benzodiazepine sensitivity in patients with panic disorder: comparison with patients with obsessive-compulsive disorder and normal subjects. Am J Psychiatry 153:1444–1449
404. Marazziti D, Carlini M, Dell'osso L (2012) Treatment strategies of obsessive-compulsive disorder and panic disorder/agoraphobia. Curr Top Med Chem 12:238–253
405. Eckel LA (2011) The ovarian hormone estradiol plays a crucial role in the control of food intake in females. Physiol Behav 104:517–524
406. Fudge MA, Kavaliers M, Ossenkopp KP (2006) Allopregnanolone produces hyperphagia by reducing neophobia without altering food palatability. Eur Neuropsychopharmacol 16:272–280
407. Higgs S, Cooper SJ (1998) Antineophobic effect of the neuroactive steroid 3alpha-hydroxy-5beta-pregnan-20-one in male rats. Pharmacol Biochem Behav 60:125–131
408. Reddy DS, Kulkarni SK (1998) The role of GABA-A and mitochondrial diazepam-binding inhibitor receptors on the effects of neurosteroids on food intake in mice. Psychopharmacology (Berl) 137:391–400
409. Flood JF, Morley JE, Roberts E (1992) Memory-enhancing effects in male mice of pregnenolone and steroids metabolically derived from it. Proc Natl Acad Sci USA 89:1567–1571
410. Maurice T, Junien JL, Privat A (1997) Dehydroepiandrosterone sulfate attenuates dizocilpine-induced learning impairment in mice via sigma 1-receptors. Behav Brain Res 83:159–164
411. Darnaudery M, Koehl M, Piazza PV, Le Moal M, Mayo W (2000) Pregnenolone sulfate increases hippocampal acetylcholine release and spatial recognition. Brain Res 852:173–179
412. Flood JF, Morley JE, Roberts E (1995) Pregnenolone sulfate enhances post-training memory processes when injected in very low doses into limbic system structures: the amygdala is by far the most sensitive. Proc Natl Acad Sci USA 92:10806–10810
413. Matthews DB, Morrow AL, Tokunaga S, McDaniel JR (2002) Acute ethanol administration and acute allopregnanolone administration impair spatial memory in the Morris water task. Alcohol Clin Exp Res 26:1747–1751

414. Mayo W, Dellu F, Robel P, Cherkaoui J, Le Moal M, Baulieu EE, Simon H (1993) Infusion of neurosteroids into the nucleus basalis magnocellularis affects cognitive processes in the rat. Brain Res 607:324–328
415. Meziane H, Mathis C, Paul SM, Ungerer A (1996) The neurosteroid pregnenolone sulfate reduces learning deficits induced by scopolamine and has promnestic effects in mice performing an appetitive learning task. Psychopharmacology (Berl) 126:323–330
416. Reddy DS, Kulkarni SK (1998) The effects of neurosteroids on acquisition and retention of a modified passive-avoidance learning task in mice. Brain Res 791:108–116
417. Robel P, Young J, Corpechot C, Mayo W, Perche F, Haug M, Simon H, Baulieu EE (1995) Biosynthesis and assay of neurosteroids in rats and mice: functional correlates. J Steroid Biochem Mol Biol 53:355–360
418. Ladurelle N, Eychenne B, Denton D, Blair-West J, Schumacher M, Robel P, Baulieu E (2000) Prolonged intracerebroventricular infusion of neurosteroids affects cognitive performances in the mouse. Brain Res 858:371–379
419. Melchior CL, Ritzmann RF (1996) Neurosteroids block the memory-impairing effects of ethanol in mice. Pharmacol Biochem Behav 53:51–56
420. Roberts E, Bologa L, Flood JF, Smith GE (1987) Effects of dehydroepiandrosterone and its sulfate on brain tissue in culture and on memory in mice. Brain Res 406:357–362
421. Flood JF, Smith GE, Roberts E (1988) Dehydroepiandrosterone and its sulfate enhance memory retention in mice. Brain Res 447:269–278
422. Tanaka M, Sokabe M (2012) Continuous de novo synthesis of neurosteroids is required for normal synaptic transmission and plasticity in the dentate gyrus of the rat hippocampus. Neuropharmacology 62:2373–2387
423. Pallares M, Darnaudery M, Day J, Le Moal M, Mayo W (1998) The neurosteroid pregnenolone sulfate infused into the nucleus basalis increases both acetylcholine release in the frontal cortex or amygdala and spatial memory. Neuroscience 87:551–558
424. Darnaudery M, Koehl M, Pallares M, Le Moal M, Mayo W (1998) The neurosteroid pregnenolone sulfate increases cortical acetylcholine release: a microdialysis study in freely moving rats. J Neurochem 71:2018–2022
425. Rhodes ME, Li PK, Flood JF, Johnson DA (1996) Enhancement of hippocampal acetylcholine release by the neurosteroid dehydroepiandrosterone sulfate: an in vivo microdialysis study. Brain Res 733:284–286
426. Rhodes ME, Li PK, Burke AM, Johnson DA (1997) Enhanced plasma DHEAS, brain acetylcholine and memory mediated by steroid sulfatase inhibition. Brain Res 773:28–32
427. Isaacson RL, Varner JA, Baars JM, de Wied D (1995) The effects of pregnenolone sulfate and ethylestrenol on retention of a passive avoidance task. Brain Res 689:79–84
428. Martin-Garcia E, Pallares M (2008) A post-training intrahippocampal anxiogenic dose of the neurosteroid pregnenolone sulfate impairs passive avoidance retention. Exp Brain Res 191:123–131
429. Mathis C, Paul SM, Crawley JN (1994) The neurosteroid pregnenolone sulfate blocks NMDA antagonist-induced deficits in a passive avoidance memory task. Psychopharmacology (Berl) 116:201–206
430. Mayo W, George O, Darbra S, Bouyer JJ, Vallee M, Darnaudery M, Pallares M, Lemaire-Mayo V, Le Moal M, Piazza PV, Abrous N (2003) Individual differences in cognitive aging: implication of pregnenolone sulfate. Prog Neurobiol 71:43–48
431. Migues PV, Johnston AN, Rose SP (2002) Dehydroepiandrosterone and its sulphate enhance memory retention in day-old chicks. Neuroscience 109:243–251
432. Vallee M, Shen W, Heinrichs SC, Zorumski CF, Covey DF, Koob GF, Purdy RH (2001) Steroid structure and pharmacological properties determine the anti-amnesic effects of pregnenolone sulphate in the passive avoidance task in rats. Eur J Neurosci 14:2003–2010
433. Reddy DS, Kulkarni SK (1998) Possible role of nitric oxide in the nootropic and antiamnesic effects of neurosteroids on aging- and dizocilpine-induced learning impairment. Brain Res 799:215–229

434. Petit GH, Tobin C, Krishnan K, Moricard Y, Covey DF, Rondi-Reig L, Akwa Y (2011) Pregnenolone sulfate and its enantiomer: differential modulation of memory in a spatial discrimination task using forebrain NMDA receptor deficient mice. Eur Neuropsychopharmacol 21:211–215
435. Lhullier FL, Nicolaidis R, Riera NG, Cipriani F, Junqueira D, Dahm KC, Brusque AM, Souza DO (2004) Dehydroepiandrosterone increases synaptosomal glutamate release and improves the performance in inhibitory avoidance task. Pharmacol Biochem Behav 77:601–606
436. Urani A, Privat A, Maurice T (1998) The modulation by neurosteroids of the scopolamine-induced learning impairment in mice involves an interaction with sigma1 (sigma1) receptors. Brain Res 799:64–77
437. Maurice T, Phan VL, Urani A, Guillemain I (2001) Differential involvement of the sigma(1) (sigma(1)) receptor in the anti-amnesic effect of neuroactive steroids, as demonstrated using an in vivo antisense strategy in the mouse. Br J Pharmacol 134:1731–1741
438. Zou LB, Yamada K, Sasa M, Nakata Y, Nabeshima T (2000) Effects of sigma(1) receptor agonist SA4503 and neuroactive steroids on performance in a radial arm maze task in rats. Neuropharmacology 39:1617–1627
439. Xu Y, Tanaka M, Chen L, Sokabe M (2012) DHEAS induces short-term potentiation via the activation of a metabotropic glutamate receptor in the rat hippocampus. Hippocampus 22:707–722
440. Milad MR, Igoe SA, Lebron-Milad K, Novales JE (2009) Estrous cycle phase and gonadal hormones influence conditioned fear extinction. Neuroscience 164:887–895
441. Workman JL, Barha CK, Galea LA (2012) Endocrine substrates of cognitive and affective changes during pregnancy and postpartum. Behav Neurosci 126:54–72
442. Foy MR (2011) Ovarian hormones, aging and stress on hippocampal synaptic plasticity. Neurobiol Learn Mem 95:134–144
443. Foy MR, Baudry M, Foy JG, Thompson RF (2008) 17beta-estradiol modifies stress-induced and age-related changes in hippocampal synaptic plasticity. Behav Neurosci 122:301–309
444. Leranth C, Shanabrough M, Horvath TL (2000) Hormonal regulation of hippocampal spine synapse density involves subcortical mediation. Neuroscience 101:349–356
445. Prange-Kiel J, Rune GM (2006) Direct and indirect effects of estrogen on rat hippocampus. Neuroscience 138:765–772
446. Prange-Kiel J, Rune GM, Leranth C (2004) Median raphe mediates estrogenic effects to the hippocampus in female rats. Eur J Neurosci 19:309–317
447. Prange-Kiel J, Wehrenberg U, Jarry H, Rune GM (2003) Para/autocrine regulation of estrogen receptors in hippocampal neurons. Hippocampus 13:226–234
448. Wehrenberg U, Prange-Kiel J, Rune GM (2001) Steroidogenic factor-1 expression in marmoset and rat hippocampus: co-localization with StAR and aromatase. J Neurochem 76:1879–1886
449. Hojo Y, Hattori TA, Enami T, Furukawa A, Suzuki K, Ishii HT, Mukai H, Morrison JH, Janssen WG, Kominami S, Harada N, Kimoto T, Kawato S (2004) Adult male rat hippocampus synthesizes estradiol from pregnenolone by cytochromes P45017alpha and P450 aromatase localized in neurons. Proc Natl Acad Sci USA 101:865–870
450. Prange-Kiel J, Jarry H, Schoen M, Kohlmann P, Lohse C, Zhou L, Rune GM (2008) Gonadotropin-releasing hormone regulates spine density via its regulatory role in hippocampal estrogen synthesis. J Cell Biol 180:417–426
451. Day M, Sung A, Logue S, Bowlby M, Arias R (2005) Beta estrogen receptor knockout (BERKO) mice present attenuated hippocampal CA1 long-term potentiation and related memory deficits in contextual fear conditioning. Behav Brain Res 164:128–131
452. Fugger HN, Foster TC, Gustafsson J, Rissman EF (2000) Novel effects of estradiol and estrogen receptor alpha and beta on cognitive function. Brain Res 883:258–264
453. Liu F, Day M, Muniz LC, Bitran D, Arias R, Revilla-Sanchez R, Grauer S, Zhang G, Kelley C, Pulito V, Sung A, Mervis RF, Navarra R, Hirst WD, Reinhart PH, Marquis KL, Moss SJ, Pangalos MN, Brandon NJ (2008) Activation of estrogen receptor-beta regulates hippocampal synaptic plasticity and improves memory. Nat Neurosci 11:334–343

454. Rissman EF, Heck AL, Leonard JE, Shupnik MA, Gustafsson JA (2002) Disruption of estrogen receptor beta gene impairs spatial learning in female mice. Proc Natl Acad Sci USA 99:3996–4001
455. Dumitriu D, Rapp PR, McEwen BS, Morrison JH (2010) Estrogen and the aging brain: an elixir for the weary cortical network. Ann N Y Acad Sci 1204:104–112
456. Tsurugizawa T, Mukai H, Tanabe N, Murakami G, Hojo Y, Kominami S, Mitsuhashi K, Komatsuzaki Y, Morrison JH, Janssen WG, Kimoto T, Kawato S (2005) Estrogen induces rapid decrease in dendritic thorns of CA3 pyramidal neurons in adult male rat hippocampus. Biochem Biophys Res Commun 337:1345–1352
457. Gould E, Woolley CS, Frankfurt M, McEwen BS (1990) Gonadal steroids regulate dendritic spine density in hippocampal pyramidal cells in adulthood. J Neurosci 10:1286–1291
458. Murphy DD, Segal M (1996) Regulation of dendritic spine density in cultured rat hippocampal neurons by steroid hormones. J Neurosci 16:4059–4068
459. Prange-Kiel J, Fester L, Zhou L, Lauke H, Carretero J, Rune GM (2006) Inhibition of hippocampal estrogen synthesis causes region-specific downregulation of synaptic protein expression in hippocampal neurons. Hippocampus 16:464–471
460. von Schassen C, Fester L, Prange-Kiel J, Lohse C, Huber C, Bottner M, Rune GM (2006) Oestrogen synthesis in the hippocampus: role in axon outgrowth. J Neuroendocrinol 18:847–856
461. Fester L, Ribeiro-Gouveia V, Prange-Kiel J, von Schassen C, Bottner M, Jarry H, Rune GM (2006) Proliferation and apoptosis of hippocampal granule cells require local oestrogen synthesis. J Neurochem 97:1136–1144
462. Rudick CN, Woolley CS (2003) Selective estrogen receptor modulators regulate phasic activation of hippocampal CA1 pyramidal cells by estrogen. Endocrinology 144:179–187
463. Rudick CN, Woolley CS (2000) Estradiol induces a phasic Fos response in the hippocampal CA1 and CA3 regions of adult female rats. Hippocampus 10:274–283
464. Rudick CN, Woolley CS (2001) Estrogen regulates functional inhibition of hippocampal CA1 pyramidal cells in the adult female rat. J Neurosci 21:6532–6543
465. Murphy DD, Cole NB, Greenberger V, Segal M (1998) Estradiol increases dendritic spine density by reducing GABA neurotransmission in hippocampal neurons. J Neurosci 18:2550–2559
466. Huang GZ, Woolley CS (2012) Estradiol acutely suppresses inhibition in the hippocampus through a sex-specific endocannabinoid and mGluR-dependent mechanism. Neuron 74:801–808
467. Wong M, Moss RL (1992) Long-term and short-term electrophysiological effects of estrogen on the synaptic properties of hippocampal CA1 neurons. J Neurosci 12:3217–3225
468. Zadran S, Qin Q, Bi X, Zadran H, Kim Y, Foy MR, Thompson R, Baudry M (2009) 17-Beta-estradiol increases neuronal excitability through MAP kinase-induced calpain activation. Proc Natl Acad Sci USA 106:21936–21941
469. Woolley CS, McEwen BS (1994) Estradiol regulates hippocampal dendritic spine density via an *N*-methyl-D-aspartate receptor-dependent mechanism. J Neurosci 14:7680–7687
470. Murakami G, Tsurugizawa T, Hatanaka Y, Komatsuzaki Y, Tanabe N, Mukai H, Hojo Y, Kominami S, Yamazaki T, Kimoto T, Kawato S (2006) Comparison between basal and apical dendritic spines in estrogen-induced rapid spinogenesis of CA1 principal neurons in the adult hippocampus. Biochem Biophys Res Commun 351:553–558
471. Nanfaro F, Cabrera R, Bazzocchini V, Laconi M, Yunes R (2010) Pregnenolone sulfate infused in lateral septum of male rats impairs novel object recognition memory. Pharmacol Rep 62:265–272
472. Kask K, Backstrom T, Nilsson LG, Sundstrom-Poromaa I (2008) Allopregnanolone impairs episodic memory in healthy women. Psychopharmacology (Berl) 199:161–168
473. Johansson IM, Birzniece V, Lindblad C, Olsson T, Backstrom T (2002) Allopregnanolone inhibits learning in the Morris water maze. Brain Res 934:125–131
474. Murphy DD, Segal M (2000) Progesterone prevents estradiol-induced dendritic spine formation in cultured hippocampal neurons. Neuroendocrinology 72:133–143

475. Wiltgen BJ, Sanders MJ, Ferguson C, Homanics GE, Fanselow MS (2005) Trace fear conditioning is enhanced in mice lacking the delta subunit of the GABAA receptor. Learn Mem 12:327–333
476. Galea LA, Spritzer MD, Barker JM, Pawluski JL (2006) Gonadal hormone modulation of hippocampal neurogenesis in the adult. Hippocampus 16:225–232
477. Shen H, Sabaliauskas N, Sherpa A, Fenton AA, Stelzer A, Aoki C, Smith SS (2010) A critical role for alpha4betadelta GABAA receptors in shaping learning deficits at puberty in mice. Science 327:1515–1518
478. Maurice T, Privat A (1997) SA4503, a novel cognitive enhancer with sigma1 receptor agonist properties, facilitates NMDA receptor-dependent learning in mice. Eur J Pharmacol 328:9–18
479. Frye CA, Walf AA (2008) Progesterone to ovariectomized mice enhances cognitive performance in the spontaneous alternation, object recognition, but not placement, water maze, and contextual and cued conditioned fear tasks. Neurobiol Learn Mem 90:171–177
480. Vallee M, Mayo W, Darnaudery M, Corpechot C, Young J, Koehl M, Le Moal M, Baulieu EE, Robel P, Simon H (1997) Neurosteroids: deficient cognitive performance in aged rats depends on low pregnenolone sulfate levels in the hippocampus. Proc Natl Acad Sci USA 94:14865–14870
481. Akwa Y, Ladurelle N, Covey DF, Baulieu EE (2001) The synthetic enantiomer of pregnenolone sulfate is very active on memory in rats and mice, even more so than its physiological neurosteroid counterpart: distinct mechanisms? Proc Natl Acad Sci USA 98:14033–14037
482. Flood JF, Roberts E (1988) Dehydroepiandrosterone sulfate improves memory in aging mice. Brain Res 448:178–181
483. Racchi M, Govoni S, Solerte SB, Galli CL, Corsini E (2001) Dehydroepiandrosterone and the relationship with aging and memory: a possible link with protein kinase C functional machinery. Brain Res Brain Res Rev 37:287–293
484. Frye CA, Rhodes ME, Dudek B (2005) Estradiol to aged female or male mice improves learning in inhibitory avoidance and water maze tasks. Brain Res 1036:101–108
485. Brinton RD (2001) Cellular and molecular mechanisms of estrogen regulation of memory function and neuroprotection against Alzheimer's disease: recent insights and remaining challenges. Learn Mem 8:121–133
486. Bernardi F, Salvestroni C, Casarosa E, Nappi RE, Lanzone A, Luisi S, Purdy RH, Petraglia F, Genazzani AR (1998) Aging is associated with changes in allopregnanolone concentrations in brain, endocrine glands and serum in male rats. Eur J Endocrinol 138:316–321
487. George O, Vallee M, Vitiello S, Le Moal M, Piazza PV, Mayo W (2010) Low brain allopregnanolone levels mediate flattened circadian activity associated with memory impairments in aged rats. Biol Psychiatry 68:956–963
488. Frye CA, Duffy CK, Walf AA (2007) Estrogens and progestins enhance spatial learning of intact and ovariectomized rats in the object placement task. Neurobiol Learn Mem 88:208–216
489. Escudero C, Casas S, Giuliani F, Bazzocchini V, Garcia S, Yunes R, Cabrera R (2012) Allopregnanolone prevents memory impairment: effect on mRNA expression and enzymatic activity of hippocampal 3-alpha hydroxysteroid oxide-reductase. Brain Res Bull 87:280–285
490. Frye CA, Walf AA (2008) Effects of progesterone administration and APPswe + PSEN1Deltae9 mutation for cognitive performance of mid-aged mice. Neurobiol Learn Mem 89:17–26
491. Singh C, Liu L, Wang JM, Irwin RW, Yao J, Chen S, Henry S, Thompson RF, Brinton RD (2012) Allopregnanolone restores hippocampal-dependent learning and memory and neural progenitor survival in aging 3xTgAD and nonTg mice. Neurobiol Aging 33:1493–1506
492. Yau JL, Noble J, Graham M, Seckl JR (2006) Central administration of a cytochrome P450-7B product 7 alpha-hydroxypregnenolone improves spatial memory retention in cognitively impaired aged rats. J Neurosci 26:11034–11040
493. Tsutsui K, Haraguchi S, Inoue K, Miyabara H, Suzuki S, Ogura Y, Koyama T, Matsunaga M, Vaudry H (2009) Identification, biosynthesis, and function of 7alpha-hydroxypregnenolone, a new key neurosteroid controlling locomotor activity, in nonmammalian vertebrates. Ann N Y Acad Sci 1163:308–315

494. Duff SJ, Hampson E (2000) A beneficial effect of estrogen on working memory in postmenopausal women taking hormone replacement therapy. Horm Behav 38:262–276
495. Keenan PA, Ezzat WH, Ginsburg K, Moore GJ (2001) Prefrontal cortex as the site of estrogen's effect on cognition. Psychoneuroendocrinology 26:577–590
496. Dumas JA, Kutz AM, Naylor MR, Johnson JV, Newhouse PA (2010) Increased memory load-related frontal activation after estradiol treatment in postmenopausal women. Horm Behav 58:929–935
497. Brinton RD (2005) Investigative models for determining hormone therapy-induced outcomes in brain: evidence in support of a healthy cell bias of estrogen action. Ann N Y Acad Sci 1052:57–74
498. Daniel JM, Bohacek J (2010) The critical period hypothesis of estrogen effects on cognition: insights from basic research. Biochim Biophys Acta 1800:1068–1076
499. Maki PM (2005) A systematic review of clinical trials of hormone therapy on cognitive function: effects of age at initiation and progestin use. Ann N Y Acad Sci 1052:182–197
500. Chisholm NC, Juraska JM (2012) Long-term replacement of estrogen in combination with medroxyprogesterone acetate improves acquisition of an alternation task in middle-aged female rats. Behav Neurosci 126:128–136
501. Gibbs RB (2000) Long-term treatment with estrogen and progesterone enhances acquisition of a spatial memory task by ovariectomized aged rats. Neurobiol Aging 21:107–116
502. Bimonte-Nelson HA, Francis KR, Umphlet CD, Granholm AC (2006) Progesterone reverses the spatial memory enhancements initiated by tonic and cyclic oestrogen therapy in middle-aged ovariectomized female rats. Eur J Neurosci 24:229–242
503. Sherwin BB, Grigorova M (2011) Differential effects of estrogen and micronized progesterone or medroxyprogesterone acetate on cognition in postmenopausal women. Fertil Steril 96:399–403
504. Cooke BM, Breedlove SM, Jordan CL (2003) Both estrogen receptors and androgen receptors contribute to testosterone-induced changes in the morphology of the medial amygdala and sexual arousal in male rats. Horm Behav 43:336–346
505. Kudwa AE, Michopoulos V, Gatewood JD, Rissman EF (2006) Roles of estrogen receptors alpha and beta in differentiation of mouse sexual behavior. Neuroscience 138:921–928
506. Sato T, Matsumoto T, Kawano H, Watanabe T, Uematsu Y, Sekine K, Fukuda T, Aihara K, Krust A, Yamada T, Nakamichi Y, Yamamoto Y, Nakamura T, Yoshimura K, Yoshizawa T, Metzger D, Chambon P, Kato S (2004) Brain masculinization requires androgen receptor function. Proc Natl Acad Sci USA 101:1673–1678
507. Merkx J (1984) Effect of castration and subsequent substitution with testosterone, dihydrotestosterone and oestradiol on sexual preference behaviour in the male rat. Behav Brain Res 11:59–65
508. King SR, Lamb DJ (2006) Why we lose interest in sex: do neurosteroids play a role? Sex Reprod Menopause 4:20–23
509. Frye CA, Sumida K, Zimmerberg B, Brunelli SA (2006) Rats bred for high versus low anxiety responses neonatally demonstrate increases in lordosis, pacing behavior, and midbrain 3 alpha, 5 alpha-THP levels as adults. Behav Neurosci 120:281–289
510. Sleiter N, Pang Y, Park C, Horton TH, Dong J, Thomas P, Levine JE (2009) Progesterone receptor A (PRA) and PRB-independent effects of progesterone on gonadotropin-releasing hormone release. Endocrinology 150:3833–3844
511. Chappell PE, Schneider JS, Kim P, Xu M, Lydon JP, O'Malley BW, Levine JE (1999) Absence of gonadotropin surges and gonadotropin-releasing hormone self-priming in ovariectomized (OVX), estrogen (E2)-treated, progesterone receptor knockout (PRKO) mice. Endocrinology 140:3653–3658
512. Chappell PE, Levine JE (2000) Stimulation of gonadotropin-releasing hormone surges by estrogen. I. Role of hypothalamic progesterone receptors. Endocrinology 141:1477–1485
513. White MM, Sheffer I, Teeter J, Apostolakis EM (2007) Hypothalamic progesterone receptor-A mediates gonadotropin surges, self priming and receptivity in estrogen-primed female mice. J Mol Endocrinol 38:35–50

514. Kuo J, Micevych P (2012) Neurosteroids, trigger of the LH surge. J Steroid Biochem Mol Biol 131:57–65
515. Micevych P, Sinchak K, Mills RH, Tao L, LaPolt P, Lu JK (2003) The luteinizing hormone surge is preceded by an estrogen-induced increase of hypothalamic progesterone in ovariectomized and adrenalectomized rats. Neuroendocrinology 78:29–35
516. Soma KK, Sinchak K, Lakhter A, Schlinger BA, Micevych PE (2005) Neurosteroids and female reproduction: estrogen increases 3beta-HSD mRNA and activity in rat hypothalamus. Endocrinology 146:4386–4390
517. Kuo J, Hamid N, Bondar G, Prossnitz ER, Micevych P (2010) Membrane estrogen receptors stimulate intracellular calcium release and progesterone synthesis in hypothalamic astrocytes. J Neurosci 30:12950–12957
518. Bondar G, Kuo J, Hamid N, Micevych P (2009) Estradiol-induced estrogen receptor-alpha trafficking. J Neurosci 29:15323–15330
519. Micevych PE, Chaban V, Ogi J, Dewing P, Lu JK, Sinchak K (2007) Estradiol stimulates progesterone synthesis in hypothalamic astrocyte cultures. Endocrinology 148:782–789
520. Zhao D, Duan H, Kim YC, Jefcoate CR (2005) Rodent StAR mRNA is substantially regulated by control of mRNA stability through sites in the 3′-untranslated region and through coupling to ongoing transcription. J Steroid Biochem Mol Biol 96:155–173
521. Sullivan SD, Moenter SM (2003) Neurosteroids alter gamma-aminobutyric acid postsynaptic currents in gonadotropin-releasing hormone neurons: a possible mechanism for direct steroidal control. Endocrinology 144:4366–4375
522. El-Etr M, Akwa Y, Baulieu EE, Schumacher M (2006) The neuroactive steroid pregnenolone sulfate stimulates the release of gonadotropin-releasing hormone from GT1-7 hypothalamic neurons, through *N*-methyl-D-aspartate receptors. Endocrinology 147:2737–2743
523. Wiebe JP, Wood PH (1987) Selective suppression of follicle-stimulating hormone by 3 alpha-hydroxy-4-pregnen-20-one, a steroid found in Sertoli cells. Endocrinology 120:2259–2264
524. Wiebe JP, Boushy D, Wolfe M (1997) Synthesis, metabolism and levels of the neuroactive steroid, 3alpha-hydroxy-4-pregnen-20-one (3alphaHP), in rat pituitaries. Brain Res 764:158–166
525. Wood PH, Wiebe JP (1989) Selective suppression of follicle-stimulating hormone secretion in anterior pituitary cells by the gonadal steroid 3 alpha-hydroxy-4-pregnen-20-one. Endocrinology 125:41–48
526. Beck CA, Wolfe M, Murphy LD, Wiebe JP (1997) Acute, nongenomic actions of the neuroactive gonadal steroid, 3 alpha-hydroxy-4-pregnen-20-one (3 alpha HP), on FSH release in perifused rat anterior pituitary cells. Endocrine 6:221–229
527. Dhanvantari S, Wiebe JP (1994) Suppression of follicle-stimulating hormone by the gonadal- and neurosteroid 3 alpha-hydroxy-4-pregnen-20-one involves actions at the level of the gonadotrope membrane/calcium channel. Endocrinology 134:371–376
528. Wiebe JP, Dhanvantari S, Watson PH, Huang Y (1994) Suppression in gonadotropes of gonadotropin-releasing hormone-stimulated follicle-stimulating hormone release by the gonadal- and neurosteroid 3 alpha-hydroxy-4-pregnen-20-one involves cytosolic calcium. Endocrinology 134:377–382
529. Timby E, Hedstrom H, Backstrom T, Sundstrom-Poromaa I, Nyberg S, Bixo M (2011) Allopregnanolone, a GABAA receptor agonist, decreases gonadotropin levels in women. A preliminary study. Gynecol Endocrinol 27:1087–1093
530. Dewing P, Boulware MI, Sinchak K, Christensen A, Mermelstein PG, Micevych P (2007) Membrane estrogen receptor-alpha interactions with metabotropic glutamate receptor 1a modulate female sexual receptivity in rats. J Neurosci 27:9294–9300
531. Kow LM, Pfaff DW (2004) The membrane actions of estrogens can potentiate their lordosis behavior-facilitating genomic actions. Proc Natl Acad Sci USA 101:12354–12357
532. Micevych P, Soma KK, Sinchak K (2008) Neuroprogesterone: key to estrogen positive feedback? Brain Res Rev 57:470–480
533. Pfaff D, Schwartz-Giblin S (1998) Cellular mechanisms of female reproductive behaviors. In: Knobil E, Neill J, Ewing L, Greenwald G, Markett C, Pfaff D (eds) The physiology of reproduction, 1st edn. Raven Press, New York, pp 1487–1568

534. Krebs CJ, Jarvis ED, Chan J, Lydon JP, Ogawa S, Pfaff DW (2000) A membrane-associated progesterone-binding protein, 25-Dx, is regulated by progesterone in brain regions involved in female reproductive behaviors. Proc Natl Acad Sci USA 97:12816–12821
535. Liu B, Arbogast LA (2009) Gene expression profiles of intracellular and membrane progesterone receptor isoforms in the mediobasal hypothalamus during pro-oestrus. J Neuroendocrinol 21:993–1000
536. Frye CA, Bayon LE (1999) Mating stimuli influence endogenous variations in the neurosteroids 3alpha,5alpha-THP and 3alpha-Diol. J Neuroendocrinol 11:839–847
537. Frye CA, Rhodes ME (2006) Progestin concentrations are increased following paced mating in midbrain, hippocampus, diencephalon, and cortex of rats in behavioral estrus, but only in midbrain of diestrous rats. Neuroendocrinology 83:336–347
538. Hassell J, Miryala CS, Hiegel C, Uphouse L (2011) Mechanisms responsible for progesterone's protection against lordosis-inhibiting effects of restraint I. Role of progesterone receptors. Horm Behav 60:219–225
539. Pazol K, Northcutt KV, Wilson ME, Wallen K (2006) Medroxyprogesterone acetate acutely facilitates and sequentially inhibits sexual behavior in female rats. Horm Behav 49:105–113
540. Kaunitz AM (2001) Injectable long-acting contraceptives. Clin Obstet Gynecol 44:73–91
541. Frye CA (2001) The role of neurosteroids and nongenomic effects of progestins in the ventral tegmental area in mediating sexual receptivity of rodents. Horm Behav 40:226–233
542. Frye CA (2001) The role of neurosteroids and non-genomic effects of progestins and androgens in mediating sexual receptivity of rodents. Brain Res Brain Res Rev 37:201–222
543. Frye CA, Paris JJ, Rhodes ME (2007) Engaging in paced mating, but neither exploratory, anti-anxiety, nor social behavior, increases 5alpha-reduced progestin concentrations in midbrain, hippocampus, striatum, and cortex. Reproduction 133:663–674
544. Frye CA, Gardiner SG (1996) Progestins can have a membrane-mediated action in rat midbrain for facilitation of sexual receptivity. Horm Behav 30:682–691
545. Frye CA, DeBold JF (1993) 3 alpha-OH-DHP and 5 alpha-THDOC implants to the ventral tegmental area facilitate sexual receptivity in hamsters after progesterone priming to the ventral medial hypothalamus. Brain Res 612:130–137
546. Petralia SM, Jahagirdar V, Frye CA (2005) Inhibiting biosynthesis and/or metabolism of progestins in the ventral tegmental area attenuates lordosis of rats in behavioural oestrus. J Neuroendocrinol 17:545–552
547. Frye CA, Vongher JM (2001) Ventral tegmental area infusions of inhibitors of the biosynthesis and metabolism of 3alpha,5alpha-THP attenuate lordosis of hormone-primed and behavioural oestrous rats and hamsters. J Neuroendocrinol 13:1076–1086
548. Laconi MR, Cabrera RJ (2002) Effect of centrally injected allopregnanolone on sexual receptivity, luteinizing hormone release, hypothalamic dopamine turnover, and release in female rats. Endocrine 17:77–83
549. Petralia SM, Frye CA (2004) In the ventral tegmental area, G-proteins and cAMP mediate the neurosteroid 3alpha,5alpha-THP's actions at dopamine type 1 receptors for lordosis of rats. Neuroendocrinology 80:233–243
550. Frye CA, van Keuren KR, Erskine MS (1996) Behavioral effects of 3 alpha-androstanediol. I: Modulation of sexual receptivity and promotion of GABA-stimulated chloride flux. Behav Brain Res 79:109–118
551. Frye CA, Duncan JE, Basham M, Erskine MS (1996) Behavioral effects of 3 alpha-androstanediol. II: hypothalamic and preoptic area actions via a GABAergic mechanism. Behav Brain Res 79:119–130
552. Frye CA, van Keuren KR, Rao PN, Erskine MS (1996) Progesterone and 3 alpha-androstanediol conjugated to bovine serum albumin affects estrous behavior when applied to the MBH and POA. Behav Neurosci 110:603–612
553. Frye CA, Sumida K, Lydon JP, O'Malley BW, Pfaff DW (2006) Mid-aged and aged wild-type and progestin receptor knockout (PRKO) mice demonstrate rapid progesterone and 3alpha,5alpha-THP-facilitated lordosis. Psychopharmacology (Berl) 185:423–432

554. McCarthy MM, Felzenberg E, Robbins A, Pfaff DW, Schwartz-Giblin S (1995) Infusions of diazepam and allopregnanolone into the midbrain central gray facilitate open-field behavior and sexual receptivity in female rats. Horm Behav 29:279–295
555. Avis NE, Zhao X, Johannes CB, Ory M, Brockwell S, Greendale GA (2005) Correlates of sexual function among multi-ethnic middle-aged women: results from the Study of Women's Health Across the Nation (SWAN). Menopause 12:385–398
556. Genazzani AR, Stomati M, Valentino V, Pluchino N, Pot E, Casarosa E, Merlini S, Giannini A, Luisi M (2011) Effect of 1-year, low-dose DHEA therapy on climacteric symptoms and female sexuality. Climacteric 14:661–668
557. Kuo J, Hamid N, Bondar G, Dewing P, Clarkson J, Micevych P (2010) Sex differences in hypothalamic astrocyte response to estradiol stimulation. Biol Sex Differ 1:7
558. Mong JA, Glaser E, McCarthy MM (1999) Gonadal steroids promote glial differentiation and alter neuronal morphology in the developing hypothalamus in a regionally specific manner. J Neurosci 19:1464–1472
559. McCarthy MM, Arnold AP (2011) Reframing sexual differentiation of the brain. Nat Neurosci 14:677–683
560. Balthazart J, Baillien M, Cornil CA, Ball GF (2004) Preoptic aromatase modulates male sexual behavior: slow and fast mechanisms of action. Physiol Behav 83:247–270
561. Huddleston GG, Michael RP, Zumpe D, Clancy AN (2003) Estradiol in the male rat amygdala facilitates mounting but not ejaculation. Physiol Behav 79:239–246
562. Huddleston GG, Paisley JC, Clancy AN (2006) Effects of estrogen in the male rat medial amygdala: infusion of an aromatase inhibitor lowers mating and bovine serum albumin-conjugated estradiol implants do not promote mating. Neuroendocrinology 83:106–116
563. Haage D, Johansson S (1999) Neurosteroid modulation of synaptic and GABA-evoked currents in neurons from the rat medial preoptic nucleus. J Neurophysiol 82:143–151
564. Haage D, Backstrom T, Johansson S (2005) Interaction between allopregnanolone and pregnenolone sulfate in modulating GABA-mediated synaptic currents in neurons from the rat medial preoptic nucleus. Brain Res 1033:58–67
565. Uchida S, Noda E, Kakazu Y, Mizoguchi Y, Akaike N, Nabekura J (2002) Allopregnanolone enhancement of GABAergic transmission in rat medial preoptic area neurons. Am J Physiol Endocrinol Metab 283:E1257–E1265
566. Schneider JS, Burgess C, Sleiter NC, Doncarlos LL, Lydon JP, O'Malley B, Levine JE (2005) Enhanced sexual behaviors and androgen receptor immunoreactivity in the male progesterone receptor knockout mouse. Endocrinology 146:4340–4348
567. Kavaliers M, Wiebe JP, Galea LA (1994) Male preference for the odors of estrous female mice is enhanced by the neurosteroid 3 alpha-hydroxy-4-pregnen-20-one (3 alpha HP). Brain Res 646:140–144
568. Kavaliers M, Kinsella DM (1995) Male preference for the odors of estrous female mice is reduced by the neurosteroid pregnenolone sulfate. Brain Res 682:222–226
569. Young J, Corpechot C, Perche F, Eychenne B, Haug M, Baulieu EE, Robel P (1996) Neurosteroids in the mouse brain: behavioral and pharmacological effects of a 3 beta-hydroxysteroid dehydrogenase inhibitor. Steroids 61:144–149
570. Schneider JS, Stone MK, Wynne-Edwards KE, Horton TH, Lydon J, O'Malley B, Levine JE (2003) Progesterone receptors mediate male aggression toward infants. Proc Natl Acad Sci USA 100:2951–2956
571. Frye CA, Babson A, Walf AA (2006) Self-administration of 3alpha-androstanediol increases locomotion and analgesia and decreases aggressive behavior of male hamsters. Pharmacol Biochem Behav 86:415–421
572. Dong E, Matsumoto K, Uzunova V, Sugaya I, Takahata H, Nomura H, Watanabe H, Costa E, Guidotti A (2001) Brain 5alpha-dihydroprogesterone and allopregnanolone synthesis in a mouse model of protracted social isolation. Proc Natl Acad Sci USA 98:2849–2854
573. Pinna G, Costa E, Guidotti A (2005) Changes in brain testosterone and allopregnanolone biosynthesis elicit aggressive behavior. Proc Natl Acad Sci USA 102:2135–2140

574. Pinna G, Dong E, Matsumoto K, Costa E, Guidotti A (2003) In socially isolated mice, the reversal of brain allopregnanolone down-regulation mediates the anti-aggressive action of fluoxetine. Proc Natl Acad Sci USA 100:2035–2040
575. Nelson M, Pinna G (2011) S-norfluoxetine microinfused into the basolateral amygdala increases allopregnanolone levels and reduces aggression in socially isolated mice. Neuropharmacology 60:1154–1159
576. George O, Vallee M, Le Moal M, Mayo W (2006) Neurosteroids and cholinergic systems: implications for sleep and cognitive processes and potential role of age-related changes. Psychopharmacology (Berl) 186:402–413
577. Darbra S, George O, Bouyer JJ, Piazza PV, Le Moal M, Mayo W (2004) Sleep-wake states and cortical synchronization control by pregnenolone sulfate into the pedunculopontine nucleus. J Neurosci Res 76:742–747
578. Darnaudery M, Pallares M, Bouyer JJ, Le Moal M, Mayo W (1999) Infusion of neurosteroids into the rat nucleus basalis affects paradoxical sleep in accordance with their memory modulating properties. Neuroscience 92:583–588
579. Darnaudery M, Bouyer JJ, Pallares M, Le Moal M, Mayo W (1999) The promnesic neurosteroid pregnenolone sulfate increases paradoxical sleep in rats. Brain Res 818:492–498
580. Friess E, Tagaya H, Trachsel L, Holsboer F, Rupprecht R (1997) Progesterone-induced changes in sleep in male subjects. Am J Physiol 272:E885–E891
581. Lancel M, Faulhaber J, Holsboer F, Rupprecht R (1996) Progesterone induces changes in sleep comparable to those of agonistic GABAA receptor modulators. Am J Physiol 271:E763–E772
582. Lancel M, Faulhaber J, Schiffelholz T, Romeo E, di Michele F, Holsboer F, Rupprecht R (1997) Allopregnanolone affects sleep in a benzodiazepine-like fashion. J Pharmacol Exp Ther 282:1213–1218
583. Lancel M, Faulhaber J, Holsboer F, Rupprecht R (1999) The GABA(A) receptor antagonist picrotoxin attenuates most sleep changes induced by progesterone. Psychopharmacology (Berl) 141:213–219
584. Mendelson WB, Martin JV, Perlis M, Wagner R, Majewska MD, Paul SM (1987) Sleep induction by an adrenal steroid in the rat. Psychopharmacology (Berl) 93:226–229
585. Lancel M, Cronlein TA, Muller-Preuss P, Holsboer F (1994) Pregnenolone enhances EEG delta activity during non-rapid eye movement sleep in the rat, in contrast to midazolam. Brain Res 646:85–94
586. Friess E, Trachsel L, Guldner J, Schier T, Steiger A, Holsboer F (1995) DHEA administration increases rapid eye movement sleep and EEG power in the sigma frequency range. Am J Physiol 268:E107–E113
587. Schiffelholz T, Holsboer F, Lancel M (2000) High doses of systemic DHEA-sulfate do not affect sleep structure and elicit moderate changes in non-REM sleep EEG in rats. Physiol Behav 69:399–404
588. Deurveilher S, Rusak B, Semba K (2009) Estradiol and progesterone modulate spontaneous sleep patterns and recovery from sleep deprivation in ovariectomized rats. Sleep 32:865–877
589. Wibowo E, Deurveilher S, Wassersug RJ, Semba K (2012) Estradiol treatment modulates spontaneous sleep and recovery after sleep deprivation in castrated male rats. Behav Brain Res 226:456–464
590. Selye H (1941) Anesthetic effect of steroid hormones. Proc Soc Exp Biol Med 46:116–121
591. Reddy DS, Apanites LA (2005) Anesthetic effects of progesterone are undiminished in progesterone receptor knockout mice. Brain Res 1033:96–101
592. Ren J, Greer JJ (2006) Neurosteroid modulation of respiratory rhythm in rats during the perinatal period. J Physiol 574:535–546
593. Hansen SL, Fjalland B, Jackson MB (2003) Modulation of GABAA receptors and neuropeptide secretion by the neurosteroid allopregnanolone in posterior and intermediate pituitary. Pharmacol Toxicol 93:91–97

594. Jaliffa CO, Howard S, Hoijman E, Salido E, Sarmiento MI, Arias P, Rosenstein RE (2005) Effect of neurosteroids on the retinal gabaergic system and electroretinographic activity in the golden hamster. J Neurochem 94:1666–1675
595. Saalmann YB, Morgan IG, Calford MB (2006) Neurosteroids involved in regulating inhibition in the inferior colliculus. J Neurophysiol 96:3064–3073
596. Nguyen PN, Yan EB, Castillo-Melendez M, Walker DW, Hirst JJ (2004) Increased allopregnanolone levels in the fetal sheep brain following umbilical cord occlusion. J Physiol 560:593–602
597. Billiards SS, Nguyen PN, Scheerlinck JP, Phillips DJ, Canny BJ, Walker DW, Hirst JJ (2006) Hypoxia potentiates endotoxin-induced allopregnanolone concentrations in the newborn brain. Biol Neonate 90:258–267
598. Ganter S, Northoff H, Mannel D, Gebicke-Harter PJ (1992) Growth control of cultured microglia. J Neurosci Res 33:218–230
599. Roof RL, Duvdevani R, Braswell L, Stein DG (1994) Progesterone facilitates cognitive recovery and reduces secondary neuronal loss caused by cortical contusion injury in male rats. Exp Neurol 129:64–69
600. Roof RL, Hoffman SW, Stein DG (1997) Progesterone protects against lipid peroxidation following traumatic brain injury in rats. Mol Chem Neuropathol 31:1–11
601. Djebaili M, Guo Q, Pettus EH, Hoffman SW, Stein DG (2005) The neurosteroids progesterone and allopregnanolone reduce cell death, gliosis, and functional deficits after traumatic brain injury in rats. J Neurotrauma 22:106–118
602. Jiang N, Chopp M, Stein D, Feit H (1996) Progesterone is neuroprotective after transient middle cerebral artery occlusion in male rats. Brain Res 735:101–107
603. Betz AL, Coester HC (1990) Effect of steroids on edema and sodium uptake of the brain during focal ischemia in rats. Stroke 21:1199–1204
604. Ishrat T, Sayeed I, Atif F, Hua F, Stein DG (2010) Progesterone and allopregnanolone attenuate blood–brain barrier dysfunction following permanent focal ischemia by regulating the expression of matrix metalloproteinases. Exp Neurol 226:183–190
605. Gibson CL, Constantin D, Prior MJ, Bath PM, Murphy SP (2005) Progesterone suppresses the inflammatory response and nitric oxide synthase-2 expression following cerebral ischemia. Exp Neurol 193:522–530
606. Stein DG (2011) Is progesterone a worthy candidate as a novel therapy for traumatic brain injury? Dialogues Clin Neurosci 13:352–359
607. Doonan F, O'Driscoll C, Kenna P, Cotter TG (2011) Enhancing survival of photoreceptor cells in vivo using the synthetic progestin Norgestrel. J Neurochem 118:915–927
608. Ciriza I, Azcoitia I, Garcia-Segura LM (2004) Reduced progesterone metabolites protect rat hippocampal neurones from kainic acid excitotoxicity in vivo. J Neuroendocrinol 16:58–63
609. Leskiewicz M, Budziszewska B, Jaworska-Feil L, Lason W (1997) Effects of neurosteroids on kainate-induced seizures, neurotoxicity and lethality in mice. Pol J Pharmacol 49:411–417
610. Marx CE, Jarskog LF, Lauder JM, Gilmore JH, Lieberman JA, Morrow AL (2000) Neurosteroid modulation of embryonic neuronal survival in vitro following anoxia. Brain Res 871:104–112
611. Rakotomamonjy J, Levenes C, Baulieu EE, Schumacher M, Ghoumari AM (2011) Novel protective effect of mifepristone on detrimental GABAA receptor activity to immature Purkinje neurons. FASEB J 25:3999–4010
612. Ghoumari AM, Dusart I, El-Etr M, Tronche F, Sotelo C, Schumacher M, Baulieu EE (2003) Mifepristone (RU486) protects Purkinje cells from cell death in organotypic slice cultures of postnatal rat and mouse cerebellum. Proc Natl Acad Sci USA 100:7953–7958
613. Ibanez C, Shields SA, El-Etr M, Leonelli E, Magnaghi V, Li WW, Sim FJ, Baulieu EE, Melcangi RC, Schumacher M, Franklin RJ (2003) Steroids and the reversal of age-associated changes in myelination and remyelination. Prog Neurobiol 71:49–56
614. Ibanez C, Shields SA, El-Etr M, Baulieu EE, Schumacher M, Franklin RJ (2004) Systemic progesterone administration results in a partial reversal of the age-associated decline in CNS remyelination following toxin-induced demyelination in male rats. Neuropathol Appl Neurobiol 30:80–89

615. Ghoumari AM, Ibanez C, El-Etr M, Leclerc P, Eychenne B, O'Malley BW, Baulieu EE, Schumacher M (2003) Progesterone and its metabolites increase myelin basic protein expression in organotypic slice cultures of rat cerebellum. J Neurochem 86:848–859
616. Ghoumari AM, Baulieu EE, Schumacher M (2005) Progesterone increases oligodendroglial cell proliferation in rat cerebellar slice cultures. Neuroscience 135:47–58
617. Weaver CE Jr, Wu FS, Gibbs TT, Farb DH (1998) Pregnenolone sulfate exacerbates NMDA-induced death of hippocampal neurons. Brain Res 803:129–136
618. Cascio C, Guarneri R, Russo D, de Leo G, Guarneri M, Piccoli F, Guarneri P (2002) A caspase-3-dependent pathway is predominantly activated by the excitotoxin pregnenolone sulfate and requires early and late cytochrome c release and cell-specific caspase-2 activation in the retinal cell death. J Neurochem 83:1358–1371
619. Guarneri P, Russo D, Cascio C, de Leo G, Piccoli T, Sciuto V, Piccoli F, Guarneri R (1998) Pregnenolone sulfate modulates NMDA receptors, inducing and potentiating acute excitotoxicity in isolated retina. J Neurosci Res 54:787–797
620. Shirakawa H, Katsuki H, Kume T, Kaneko S, Akaike A (2005) Pregnenolone sulphate attenuates AMPA cytotoxicity on rat cortical neurons. Eur J Neurosci 21:2329–2335
621. Garcia-Segura LM, Veiga S, Sierra A, Melcangi RC, Azcoitia I (2003) Aromatase: a neuroprotective enzyme. Prog Neurobiol 71:31–41
622. Azcoitia I, Sierra A, Garcia-Segura LM (1998) Estradiol prevents kainic acid-induced neuronal loss in the rat dentate gyrus. Neuroreport 9:3075–3079
623. Wilson ME, Liu Y, Wise PM (2002) Estradiol enhances Akt activation in cortical explant cultures following neuronal injury. Brain Res Mol Brain Res 102:48–54
624. Regan RF, Guo Y (1997) Estrogens attenuate neuronal injury due to hemoglobin, chemical hypoxia, and excitatory amino acids in murine cortical cultures. Brain Res 764:133–140
625. Singer CA, Rogers KL, Strickland TM, Dorsa DM (1996) Estrogen protects primary cortical neurons from glutamate toxicity. Neurosci Lett 212:13–16
626. Zaulyanov LL, Green PS, Simpkins JW (1999) Glutamate receptor requirement for neuronal death from anoxia-reoxygenation: an in Vitro model for assessment of the neuroprotective effects of estrogens. Cell Mol Neurobiol 19:705–718
627. Sawada H, Ibi M, Kihara T, Urushitani M, Akaike A, Shimohama S (1998) Estradiol protects mesencephalic dopaminergic neurons from oxidative stress-induced neuronal death. J Neurosci Res 54:707–719
628. Cascio C, Russo D, Drago G, Galizzi G, Passantino R, Guarneri R, Guarneri P (2007) 17beta-estradiol synthesis in the adult male rat retina. Exp Eye Res 85:166–172
629. Nakazawa T, Takahashi H, Shimura M (2006) Estrogen has a neuroprotective effect on axotomized RGCs through ERK signal transduction pathway. Brain Res 1093:141–149
630. Singer CA, Rogers KL, Dorsa DM (1998) Modulation of Bcl-2 expression: a potential component of estrogen protection in NT2 neurons. Neuroreport 9:2565–2568
631. Sawada H, Ibi M, Kihara T, Urushitani M, Honda K, Nakanishi M, Akaike A, Shimohama S (2000) Mechanisms of antiapoptotic effects of estrogens in nigral dopaminergic neurons. FASEB J 14:1202–1214
632. Alexaki VI, Charalampopoulos I, Kampa M, Nifli AP, Hatzoglou A, Gravanis A, Castanas E (2006) Activation of membrane estrogen receptors induce pro-survival kinases. J Steroid Biochem Mol Biol 98:97–110
633. Aguirre C, Jayaraman A, Pike C, Baudry M (2010) Progesterone inhibits estrogen-mediated neuroprotection against excitotoxicity by down-regulating estrogen receptor-beta. J Neurochem 115:1277–1287
634. Kurata K, Takebayashi M, Morinobu S, Yamawaki S (2004) Beta-estradiol, dehydroepiandrosterone, and dehydroepiandrosterone sulfate protect against *N*-methyl-D-aspartate-induced neurotoxicity in rat hippocampal neurons by different mechanisms. J Pharmacol Exp Ther 311:237–245
635. Tiwari-Woodruff S, Morales LB, Lee R, Voskuhl RR (2007) Differential neuroprotective and antiinflammatory effects of estrogen receptor (ER)alpha and ERbeta ligand treatment. Proc Natl Acad Sci USA 104:14813–14818

636. Veiga S, Melcangi RC, Doncarlos LL, Garcia-Segura LM, Azcoitia I (2004) Sex hormones and brain aging. Exp Gerontol 39:1623–1631
637. Kretz O, Fester L, Wehrenberg U, Zhou L, Brauckmann S, Zhao S, Prange-Kiel J, Naumann T, Jarry H, Frotscher M, Rune GM (2004) Hippocampal synapses depend on hippocampal estrogen synthesis. J Neurosci 24:5913–5921
638. Saldanha CJ, Duncan KA, Walters BJ (2009) Neuroprotective actions of brain aromatase. Front Neuroendocrinol 30:106–118
639. Veiga S, Azcoitia I, Garcia-Segura LM (2005) Extragonadal synthesis of estradiol is protective against kainic acid excitotoxic damage to the hippocampus. Neuroreport 16:1599–1603
640. Dubal DB, Zhu H, Yu J, Rau SW, Shughrue PJ, Merchenthaler I, Kindy MS, Wise PM (2001) Estrogen receptor alpha, not beta, is a critical link in estradiol-mediated protection against brain injury. Proc Natl Acad Sci USA 98:1952–1957
641. Dubal DB, Rau SW, Shughrue PJ, Zhu H, Yu J, Cashion AB, Suzuki S, Gerhold LM, Bottner MB, Dubal SB, Merchanthaler I, Kindy MS, Wise PM (2006) Differential modulation of estrogen receptors (ERs) in ischemic brain injury: a role for ERalpha in estradiol-mediated protection against delayed cell death. Endocrinology 147:3076–3084
642. Garcia-Segura LM, Wozniak A, Azcoitia I, Rodriguez JR, Hutchison RE, Hutchison JB (1999) Aromatase expression by astrocytes after brain injury: implications for local estrogen formation in brain repair. Neuroscience 89:567–578
643. Chia WJ, Jenner AM, Farooqui AA, Ong WY (2008) Changes in cytochrome P450 side chain cleavage expression in the rat hippocampus after kainate injury. Exp Brain Res 186:143–149
644. Guo J, Krause DN, Horne J, Weiss JH, Li X, Duckles SP (2010) Estrogen-receptor-mediated protection of cerebral endothelial cell viability and mitochondrial function after ischemic insult in vitro. J Cereb Blood Flow Metab 30:545–554
645. Hammond J, Le Q, Goodyer C, Gelfand M, Trifiro M, LeBlanc A (2001) Testosterone-mediated neuroprotection through the androgen receptor in human primary neurons. J Neurochem 77:1319–1326
646. Ahlbom E, Prins GS, Ceccatelli S (2001) Testosterone protects cerebellar granule cells from oxidative stress-induced cell death through a receptor mediated mechanism. Brain Res 892:255–262
647. Estrada M, Uhlen P, Ehrlich BE (2006) Ca2+ oscillations induced by testosterone enhance neurite outgrowth. J Cell Sci 119:733–743
648. Estrada M, Varshney A, Ehrlich BE (2006) Elevated testosterone induces apoptosis in neuronal cells. J Biol Chem 281:25492–25501
649. Yang SH, Perez E, Cutright J, Liu R, He Z, Day AL, Simpkins JW (2002) Testosterone increases neurotoxicity of glutamate in vitro and ischemia-reperfusion injury in an animal model. J Appl Physiol 92:195–201
650. Uchida M, Palmateer JM, Herson PS, DeVries AC, Cheng J, Hurn PD (2009) Dose-dependent effects of androgens on outcome after focal cerebral ischemia in adult male mice. J Cereb Blood Flow Metab 29:1454–1462
651. Gibson CL, Coomber B, Rathbone J (2009) Is progesterone a candidate neuroprotective factor for treatment following ischemic stroke? Neuroscientist 15:324–332
652. Rosario ER, Ramsden M, Pike CJ (2006) Progestins inhibit the neuroprotective effects of estrogen in rat hippocampus. Brain Res 1099:206–210
653. Shumaker SA, Legault C, Rapp SR, Thal L, Wallace RB, Ockene JK, Hendrix SL, Jones BN III, Assaf AR, Jackson RD, Kotchen JM, Wassertheil-Smoller S, Wactawski-Wende J (2003) Estrogen plus progestin and the incidence of dementia and mild cognitive impairment in postmenopausal women: the Women's Health Initiative Memory Study: a randomized controlled trial. JAMA 289:2651–2662
654. Pike CJ, Carroll JC, Rosario ER, Barron AM (2009) Protective actions of sex steroid hormones in Alzheimer's disease. Front Neuroendocrinol 30:239–258
655. Carroll JC, Rosario ER, Villamagna A, Pike CJ (2010) Continuous and cyclic progesterone differentially interact with estradiol in the regulation of Alzheimer-like pathology in female 3xTransgenic-Alzheimer's disease mice. Endocrinology 151:2713–2722

656. Nilsen J, Brinton RD (2002) Impact of progestins on estrogen-induced neuroprotection: synergy by progesterone and 19-norprogesterone and antagonism by medroxyprogesterone acetate. Endocrinology 143:205–212
657. Irwin RW, Yao J, Ahmed SS, Hamilton RT, Cadenas E, Brinton RD (2011) Medroxyprogesterone acetate antagonizes estrogen up-regulation of brain mitochondrial function. Endocrinology 152:556–567
658. Labombarda F, Meffre D, Delespierre B, Krivokapic-Blondiaux S, Chastre A, Thomas P, Pang Y, Lydon JP, Gonzalez SL, de Nicola AF, Schumacher M, Guennoun R (2010) Membrane progesterone receptors localization in the mouse spinal cord. Neuroscience 166:94–106
659. Altman K, Gordon GG, Southren AL, Vittek J, Wilker S (1972) Induction of hepatic testosterone A-ring reductase by medroxyprogesterone acetate. Endocrinology 90:1252–1260
660. Jarrell J (1984) Studies on the developmental pattern of rat ovarian 3 alpha-hydroxysteroid dehydrogenase: inhibition of the postpubertal activity with medroxyprogesterone acetate in vivo. J Steroid Biochem 21:151–156
661. Bernardi F, Pluchino N, Pieri M, Begliuomini S, Lenzi E, Puccetti S, Casarosa E, Luisi M, Genazzani AR (2006) Progesterone and medroxyprogesterone acetate effects on central and peripheral allopregnanolone and beta-endorphin levels. Neuroendocrinology 83:348–359
662. Li Z, Cui S, Zhang Z, Zhou R, Ge Y, Sokabe M, Chen L (2009) DHEA-neuroprotection and -neurotoxicity after transient cerebral ischemia in rats. J Cereb Blood Flow Metab 29:287–296
663. Kimonides VG, Khatibi NH, Svendsen CN, Sofroniew MV, Herbert J (1998) Dehydroepiandrosterone (DHEA) and DHEA-sulfate (DHEAS) protect hippocampal neurons against excitatory amino acid-induced neurotoxicity. Proc Natl Acad Sci USA 95:1852–1857
664. Morfin R, Starka L (2001) Neurosteroid 7-hydroxylation products in the brain. Int Rev Neurobiol 46:79–95
665. Bastianetto S, Ramassamy C, Poirier J, Quirion R (1999) Dehydroepiandrosterone (DHEA) protects hippocampal cells from oxidative stress-induced damage. Brain Res Mol Brain Res 66:35–41
666. Cardounel A, Regelson W, Kalimi M (1999) Dehydroepiandrosterone protects hippocampal neurons against neurotoxin-induced cell death: mechanism of action. Proc Soc Exp Biol Med 222:145–149
667. Charalampopoulos I, Tsatsanis C, Dermitzaki E, Alexaki VI, Castanas E, Margioris AN, Gravanis A (2004) Dehydroepiandrosterone and allopregnanolone protect sympathoadrenal medulla cells against apoptosis via antiapoptotic Bcl-2 proteins. Proc Natl Acad Sci USA 101:8209–8214
668. Charalampopoulos I, Alexaki VI, Lazaridis I, Dermitzaki E, Avlonitis N, Tsatsanis C, Calogeropoulou T, Margioris AN, Castanas E, Gravanis A (2006) G protein-associated, specific membrane binding sites mediate the neuroprotective effect of dehydroepiandrosterone. FASEB J 20:577–579
669. Zhang L, Li B, Ma W, Barker JL, Chang YH, Zhao W, Rubinow DR (2002) Dehydroepiandrosterone (DHEA) and its sulfated derivative (DHEAS) regulate apoptosis during neurogenesis by triggering the Akt signaling pathway in opposing ways. Brain Res Mol Brain Res 98:58–66
670. Li Z, Zhou R, Cui S, Xie G, Cai W, Sokabe M, Chen L (2006) Dehydroepiandrosterone sulfate prevents ischemia-induced impairment of long-term potentiation in rat hippocampal CA1 by up-regulating tyrosine phosphorylation of NMDA receptor. Neuropharmacology 51:958–966
671. di Michele F, Longone P, Romeo E, Lucchetti S, Brusa L, Pierantozzi M, Bassi A, Bernardi G, Stanzione P (2003) Decreased plasma and cerebrospinal fluid content of neuroactive steroids in Parkinson's disease. Neurol Sci 24:172–173
672. Griffin LD, Gong W, Verot L, Mellon SH (2004) Niemann-Pick type C disease involves disrupted neurosteroidogenesis and responds to allopregnanolone. Nat Med 10:704–711

673. Kim HN, Lee SJ, Koh JY (2012) The neurosteroids, allopregnanolone and progesterone, induce autophagy in cultured astrocytes. Neurochem Int 60:125–133
674. Taylor LC, Puranam K, Gilmore W, Ting JP, Matsushima GK (2010) 17beta-estradiol protects male mice from cuprizone-induced demyelination and oligodendrocyte loss. Neurobiol Dis 39:127–137
675. Le Goascogne C, Eychenne B, Tonon MC, Lachapelle F, Baumann N, Robel P (2000) Neurosteroid progesterone is up-regulated in the brain of jimpy and shiverer mice. Glia 29:14–24
676. Luchetti S, Bossers K, Frajese GV, Swaab DF (2010) Neurosteroid biosynthetic pathway changes in substantia nigra and caudate nucleus in Parkinson's disease. Brain Pathol 20:945–951
677. D'Astous M, Morissette M, Tanguay B, Callier S, Di PT (2003) Dehydroepiandrosterone (DHEA) such as 17beta-estradiol prevents MPTP-induced dopamine depletion in mice. Synapse 47:10–14
678. Dluzen DE, McDermott JL, Liu B (1996) Estrogen as a neuroprotectant against MPTP-induced neurotoxicity in C57/B1 mice. Neurotoxicol Teratol 18:603–606
679. Grandbois M, Morissette M, Callier S, Di PT (2000) Ovarian steroids and raloxifene prevent MPTP-induced dopamine depletion in mice. Neuroreport 11:343–346
680. Sawada H, Ibi M, Kihara T, Honda K, Nakamizo T, Kanki R, Nakanishi M, Sakka N, Akaike A, Shimohama S (2002) Estradiol protects dopaminergic neurons in a MPP+Parkinson's disease model. Neuropharmacology 42:1056–1064
681. Al Sweidi S, Sanchez MG, Bourque M, Morissette M, Dluzen D, Di PT (2012) Oestrogen receptors and signalling pathways: implications for neuroprotective effects of sex steroids in Parkinson's disease. J Neuroendocrinol 24:48–61
682. Chao OY, Huston JP, von Bothmer A, Pum ME (2011) Chronic progesterone treatment of male rats with unilateral 6-hydroxydopamine lesion of the dorsal striatum exasperates parkinsonian symptoms. Neuroscience 196:228–236
683. Ahmad I, Lope-Piedrafita S, Bi X, Hicks C, Yao Y, Yu C, Chaitkin E, Howison CM, Weberg L, Trouard TP, Erickson RP (2005) Allopregnanolone treatment, both as a single injection or repetitively, delays demyelination and enhances survival of Niemann-Pick C mice. J Neurosci Res 82:811–821
684. Langmade SJ, Gale SE, Frolov A, Mohri I, Suzuki K, Mellon SH, Walkley SU, Covey DF, Schaffer JE, Ory DS (2006) Pregnane X receptor (PXR) activation: a mechanism for neuroprotection in a mouse model of Niemann-Pick C disease. Proc Natl Acad Sci USA 103:13807–13812
685. Rosenbaum AI, Maxfield FR (2011) Niemann-Pick type C disease: molecular mechanisms and potential therapeutic approaches. J Neurochem 116:789–795
686. Ahboucha S, Gamrani H, Baker G (2012) GABAergic neurosteroids: the "endogenous benzodiazepines" of acute liver failure. Neurochem Int 60:707–714
687. Ahboucha S, Talani G, Fanutza T, Sanna E, Biggio G, Gamrani H, Butterworth RF (2012) Reduced brain levels of DHEAS in hepatic coma patients: significance for increased GABAergic tone in hepatic encephalopathy. Neurochem Int 61:48–53
688. Marx CE, Trost WT, Shampine LJ, Stevens RD, Hulette CM, Steffens DC, Ervin JF, Butterfield MI, Blazer DG, Massing MW, Lieberman JA (2006) The neurosteroid allopregnanolone is reduced in prefrontal cortex in Alzheimer's disease. Biol Psychiatry 60:1287–1294
689. Luchetti S, Bossers K, van de Bilt S, Agrapart V, Morales RR, Frajese GV, Swaab DF (2011) Neurosteroid biosynthetic pathways changes in prefrontal cortex in Alzheimer's disease. Neurobiol Aging 32:1964–1976
690. Mackenzie SM, Dewar D, Stewart W, Fraser R, Connell JM, Davies E (2008) The transcription of steroidogenic genes in the human cerebellum and hippocampus: a comparative survey of normal and Alzheimer's tissue. J Endocrinol 196:123–130
691. Brown RC, Han Z, Cascio C, Papadopoulos V (2003) Oxidative stress-mediated DHEA formation in Alzheimer's disease pathology. Neurobiol Aging 24:57–65

692. Wang JM, Singh C, Liu L, Irwin RW, Chen S, Chung EJ, Thompson RF, Brinton RD (2010) Allopregnanolone reverses neurogenic and cognitive deficits in mouse model of Alzheimer's disease. Proc Natl Acad Sci USA 107:6498–6503
693. Chen S, Wang JM, Irwin RW, Yao J, Liu L, Brinton RD (2011) Allopregnanolone promotes regeneration and reduces beta-amyloid burden in a preclinical model of Alzheimer's disease. PLoS One 6:e24293
694. Janicki SC, Schupf N (2010) Hormonal influences on cognition and risk for Alzheimer's disease. Curr Neurol Neurosci Rep 10:359–366
695. Rosario ER, Pike CJ (2008) Androgen regulation of beta-amyloid protein and the risk of Alzheimer's disease. Brain Res Rev 57:444–453
696. Rosario ER, Carroll J, Pike CJ (2010) Testosterone regulation of Alzheimer-like neuropathology in male 3xTg-AD mice involves both estrogen and androgen pathways. Brain Res 1359:281–290
697. Rosario ER, Carroll JC, Oddo S, LaFerla FM, Pike CJ (2006) Androgens regulate the development of neuropathology in a triple transgenic mouse model of Alzheimer's disease. J Neurosci 26:13384–13389
698. Yau JL, Rasmuson S, Andrew R, Graham M, Noble J, Olsson T, Fuchs E, Lathe R, Seckl JR (2003) Dehydroepiandrosterone 7-hydroxylase CYP7B: predominant expression in primate hippocampus and reduced expression in Alzheimer's disease. Neuroscience 121:307–314
699. Zhao L, Morgan TE, Mao Z, Lin S, Cadenas E, Finch CE, Pike CJ, Mack WJ, Brinton RD (2012) Continuous versus cyclic progesterone exposure differentially regulates hippocampal gene expression and functional profiles. PLoS One 7:e31267
700. Bali N, Arimoto JM, Iwata N, Lin SW, Zhao L, Brinton RD, Morgan TE, Finch CE (2012) Differential responses of progesterone receptor membrane component-1 (Pgrmc1) and the classical progesterone receptor (Pgr) to 17beta-estradiol and progesterone in hippocampal subregions that support synaptic remodeling and neurogenesis. Endocrinology 153:759–769
701. Frye CA, Reed TA (1998) Androgenic neurosteroids: anti-seizure effects in an animal model of epilepsy. Psychoneuroendocrinology 23:385–399
702. Reddy DS (2004) Anticonvulsant activity of the testosterone-derived neurosteroid 3alpha-androstanediol. Neuroreport 15:515–518
703. Frye CA, Scalise TJ (2000) Anti-seizure effects of progesterone and 3alpha,5alpha-THP in kainic acid and perforant pathway models of epilepsy. Psychoneuroendocrinology 25:407–420
704. Pack AM, Reddy DS, Duncan S, Herzog A (2011) Neuroendocrinological aspects of epilepsy: important issues and trends in future research. Epilepsy Behav 22:94–102
705. Gasior M, Carter RB, Goldberg SR, Witkin JM (1997) Anticonvulsant and behavioral effects of neuroactive steroids alone and in conjunction with diazepam. J Pharmacol Exp Ther 282:543–553
706. Kokate TG, Cohen AL, Karp E, Rogawski MA (1996) Neuroactive steroids protect against pilocarpine- and kainic acid-induced limbic seizures and status epilepticus in mice. Neuropharmacology 35:1049–1056
707. Frye CA, Manjarrez J, Camacho-Arroyo I (2000) Infusion of 3alpha,5alpha-THP to the pontine reticular formation attenuates PTZ-induced seizures. Brain Res 881:98–102
708. Belelli D, Bolger MB, Gee KW (1989) Anticonvulsant profile of the progesterone metabolite 5 alpha-pregnan-3 alpha-ol-20-one. Eur J Pharmacol 166:325–329
709. Verleye M, Heulard I, Gillardin JM (2008) Investigation of the anticonvulsive effect of acute immobilization stress in anxious Balb/cByJ mice using GABA A-related mechanistic probes. Psychopharmacology (Berl) 197:523–534
710. Lonsdale D, Nylen K, McIntyre Burnham W (2006) The anticonvulsant effects of progesterone and its metabolites on amygdala-kindled seizures in male rats. Brain Res 1101:110–116
711. Lonsdale D, Burnham WM (2003) The anticonvulsant effects of progesterone and 5alpha-dihydroprogesterone on amygdala-kindled seizures in rats. Epilepsia 44:1494–1499
712. Reddy DS, Castaneda DC, O'Malley BW, Rogawski MA (2004) Anticonvulsant activity of progesterone and neurosteroids in progesterone receptor knockout mice. J Pharmacol Exp Ther 310:230–239

713. Kokate TG, Banks MK, Magee T, Yamaguchi S, Rogawski MA (1999) Finasteride, a 5alpha-reductase inhibitor, blocks the anticonvulsant activity of progesterone in mice. J Pharmacol Exp Ther 288:679–684
714. Biagini G, Baldelli E, Longo D, Pradelli L, Zini I, Rogawski MA, Avoli M (2006) Endogenous neurosteroids modulate epileptogenesis in a model of temporal lobe epilepsy. Exp Neurol 201:519–524
715. Leroy C, Poisbeau P, Keller AF, Nehlig A (2004) Pharmacological plasticity of GABA(A) receptors at dentate gyrus synapses in a rat model of temporal lobe epilepsy. J Physiol 557:473–487
716. Rajasekaran K, Joshi S, Sun C, Mtchedlishvilli Z, Kapur J (2010) Receptors with low affinity for neurosteroids and GABA contribute to tonic inhibition of granule cells in epileptic animals. Neurobiol Dis 40:490–501
717. Sun C, Mtchedlishvili Z, Erisir A, Kapur J (2007) Diminished neurosteroid sensitivity of synaptic inhibition and altered location of the alpha4 subunit of GABA(A) receptors in an animal model of epilepsy. J Neurosci 27:12641–12650
718. Qi JS, Yao J, Fang C, Luscher B, Chen G (2006) Downregulation of tonic GABA currents following epileptogenic stimulation of rat hippocampal cultures. J Physiol 577:579–590
719. Payne HL, Donoghue PS, Connelly WM, Hinterreiter S, Tiwari P, Ives JH, Hann V, Sieghart W, Lees G, Thompson CL (2006) Aberrant GABA(A) receptor expression in the dentate gyrus of the epileptic mutant mouse stargazer. J Neurosci 26:8600–8608
720. Pisu MG, Mostallino MC, Dore R, Mura ML, Maciocco E, Russo E, de Sarro G, Serra M (2008) Neuroactive steroids and GABAA receptor plasticity in the brain of the WAG/Rij rat, a model of absence epilepsy. J Neurochem 106:2502–2514
721. Citraro R, Russo E, Di Paola ED, Ibbadu GF, Gratteri S, Marra R, de Sarro G (2006) Effects of some neurosteroids injected into some brain areas of WAG/Rij rats, an animal model of generalized absence epilepsy. Neuropharmacology 50:1059–1071
722. Budziszewska B, van Luijtelaar G, Coenen AM, Leskiewicz M, Lason W (1999) Effects of neurosteroids on spike-wave discharges in the genetic epileptic WAG/Rij rat. Epilepsy Res 33:23–29
723. Smith SS, Shen H, Gong QH, Zhou X (2007) Neurosteroid regulation of GABA(A) receptors: focus on the alpha4 and delta subunits. Pharmacol Ther 116:58–76
724. Kerrigan JF, Shields WD, Nelson TY, Bluestone DL, Dodson WE, Bourgeois BF, Pellock JM, Morton LD, Monaghan EP (2000) Ganaxolone for treating intractable infantile spasms: a multicenter, open-label, add-on trial. Epilepsy Res 42:133–139
725. Pieribone VA, Tsai J, Soufflet C, Rey E, Shaw K, Giller E, Dulac O (2007) Clinical evaluation of ganaxolone in pediatric and adolescent patients with refractory epilepsy. Epilepsia 48:1870–1874
726. Herzog AG, Farina EL, Drislane FW, Schomer DL, Smithson SD, Fowler KM, Dworetzky BA, Bromfield EB (2010) A comparison of anastrozole and testosterone versus placebo and testosterone for treatment of sexual dysfunction in men with epilepsy and hypogonadism. Epilepsy Behav 17:264–271
727. Reddy DS, Mohan A (2011) Development and persistence of limbic epileptogenesis are impaired in mice lacking progesterone receptors. J Neurosci 31:650–658
728. Salazar P, Tapia R (2012) Allopregnanolone potentiates the glutamate-mediated seizures induced by 4-aminopyridine in Rat hippocampus in vivo. Neurochem Res 37:596–603
729. Kokate TG, Juhng KN, Kirkby RD, Llamas J, Yamaguchi S, Rogawski MA (1999) Convulsant actions of the neurosteroid pregnenolone sulfate in mice. Brain Res 831:119–124
730. Williamson J, Mtchedlishvili Z, Kapur J (2004) Characterization of the convulsant action of pregnenolone sulfate. Neuropharmacology 46:856–864
731. Maione S, Berrino L, Vitagliano S, Leyva J, Rossi F (1992) Pregnenolone sulfate increases the convulsant potency of *N*-methyl-D-aspartate in mice. Eur J Pharmacol 219:477–479
732. Reddy DS, Kulkarni SK (1998) Proconvulsant effects of neurosteroids pregnenolone sulfate and dehydroepiandrosterone sulfate in mice. Eur J Pharmacol 345:55–59

733. Woolley CS (2000) Estradiol facilitates kainic acid-induced, but not flurothyl-induced, behavioral seizure activity in adult female rats. Epilepsia 41:510–515
734. Foy MR, Xu J, Xie X, Brinton RD, Thompson RF, Berger TW (1999) 17beta-estradiol enhances NMDA receptor-mediated EPSPs and long-term potentiation. J Neurophysiol 81:925–929
735. Veliskova J (2006) The role of estrogens in seizures and epilepsy: the bad guys or the good guys? Neuroscience 138:837–844
736. Buterbaugh GG, Hudson GM (1991) Estradiol replacement to female rats facilitates dorsal hippocampal but not ventral hippocampal kindled seizure acquisition. Exp Neurol 111:55–64
737. Harden CL (2008) Hormone replacement therapy: will it affect seizure control and AED levels? Seizure 17:176–180
738. Budziszewska B, Leskiewicz M, Kubera M, Jaworska-Feil L, Kajta M, Lason W (2001) Estrone, but not 17 beta-estradiol, attenuates kainate-induced seizures and toxicity in male mice. Exp Clin Endocrinol Diabetes 109:168–173
739. Teyler TJ, Vardaris RM, Lewis D, Rawitch AB (1980) Gonadal steroids: effects on excitability of hippocampal pyramidal cells. Science 209:1017–1018
740. Weidenfeld J, Schiller H (1984) Metabolism of steroids by human brain tumors. Clin Neuropharmacol 7:395–397
741. Seol HJ, Kim JE, Wang KC, Kim SK, Seo JS, Park SH, Jung HW (2009) The pattern of gene expression and possible relation of steroidogenic genes in oligodendroglial tumors. Int J Oncol 34:181–190
742. Labombarda F, Pianos A, Liere P, Eychenne B, Gonzalez S, Cambourg A, de Nicola AF, Schumacher M, Guennoun R (2006) Injury elicited increase in spinal cord neurosteroid content analyzed by gas chromatography mass spectrometry. Endocrinology 147:1847–1859
743. Coirini H, Gouezou M, Liere P, Delespierre B, Pianos A, Eychenne B, Schumacher M, Guennoun R (2002) 3 beta-hydroxysteroid dehydrogenase expression in rat spinal cord. Neuroscience 113:883–891
744. Venard C, Boujedaini N, Belon P, Mensah-Nyagan AG, Patte-Mensah C (2008) Regulation of neurosteroid allopregnanolone biosynthesis in the rat spinal cord by glycine and the alkaloidal analogs strychnine and gelsemine. Neuroscience 153:154–161
745. Runko E, Kaprielian Z (2002) Expression of Vema in the developing mouse spinal cord and optic chiasm. J Comp Neurol 451:289–299
746. Runko E, Kaprielian Z (2004) Caenorhabditis elegans VEM-1, a novel membrane protein, regulates the guidance of ventral nerve cord-associated axons. J Neurosci 24:9015–9026
747. Thomas AJ, Nockels RP, Pan HQ, Shaffrey CI, Chopp M (1999) Progesterone is neuroprotective after acute experimental spinal cord trauma in rats. Spine 24:2134–2138
748. Guth L, Zhang Z, Roberts E (1994) Key role for pregnenolone in combination therapy that promotes recovery after spinal cord injury. Proc Natl Acad Sci USA 91:12308–12312
749. Popovich PG, Tovar CA, Wei P, Fisher L, Jakeman LB, Basso DM (2012) A reassessment of a classic neuroprotective combination therapy for spinal cord injured rats: LPS/pregnenolone/indomethacin. Exp Neurol 233:677–685
750. Fargo KN, Foecking EM, Jones KJ, Sengelaub DR (2009) Neuroprotective actions of androgens on motoneurons. Front Neuroendocrinol 30:130–141
751. Lapchak PA, Chapman DF, Nunez SY, Zivin JA (2000) Dehydroepiandrosterone sulfate is neuroprotective in a reversible spinal cord ischemia model: possible involvement of GABA(A) receptors. Stroke 31:1953–1956
752. Duchossoy Y, David S, Baulieu EE, Robel P (2011) Treatment of experimental spinal cord injury with 3beta-methoxy-pregnenolone. Brain Res 1403:57–66
753. Gonzalez Deniselle MC, Lopez-Costa JJ, Saavedra JP, Pietranera L, Gonzalez SL, Garay L, Guennoun R, Schumacher M, de Nicola AF (2002) Progesterone neuroprotection in the Wobbler mouse, a genetic model of spinal cord motor neuron disease. Neurobiol Dis 11:457–468

754. Gonzalez Deniselle MC, Garay L, Gonzalez S, Guennoun R, Schumacher M, de Nicola AF (2005) Progesterone restores retrograde labeling of cervical motoneurons in Wobbler mouse motoneuron disease. Exp Neurol 195:518–523
755. Labombarda F, Gonzalez S, Gonzalez Deniselle MC, Garay L, Guennoun R, Schumacher M, de Nicola AF (2006) Progesterone increases the expression of myelin basic protein and the number of cells showing NG2 immunostaining in the lesioned spinal cord. J Neurotrauma 23:181–192
756. Meyer M, Gonzalez Deniselle MC, Gargiulo-Monachelli G, Garay LI, Schumacher M, Guennoun R, de Nicola AF (2012) Progesterone effects on neuronal brain-derived neurotrophic factor and glial cells during progression of Wobbler mouse neurodegeneration. Neuroscience 201:267–279
757. Gargiulo MG, Meyer M, Rodriguez GE, Garay LI, Sica RE, de Nicola AF, Gonzalez Deniselle MC (2011) Endogenous progesterone is associated to amyotrophic lateral sclerosis prognostic factors. Acta Neurol Scand 123:60–67
758. Drew PD, Chavis JA, Bhatt R (2003) Sex steroid regulation of microglial cell activation: relevance to multiple sclerosis. Ann N Y Acad Sci 1007:329–334
759. Noorbakhsh F, Ellestad KK, Maingat F, Warren KG, Han MH, Steinman L, Baker GB, Power C (2011) Impaired neurosteroid synthesis in multiple sclerosis. Brain 134:2703–2721
760. Kipp M, Amor S, Krauth R, Beyer C (2012) Multiple sclerosis: neuroprotective alliance of estrogen-progesterone and gender. Front Neuroendocrinol 33:1–16
761. Garay L, Deniselle MC, Lima A, Roig P, de Nicola AF (2007) Effects of progesterone in the spinal cord of a mouse model of multiple sclerosis. J Steroid Biochem Mol Biol 107:228–237
762. Melcangi RC, Cavarretta IT, Ballabio M, Leonelli E, Schenone A, Azcoitia I, Miguel Garcia-Segura L, Magnaghi V (2005) Peripheral nerves: a target for the action of neuroactive steroids. Brain Res Brain Res Rev 48:328–338
763. Viero C, Mechaly I, Aptel H, Puech S, Valmier J, Bancel F, Dayanithi G (2006) Rapid inhibition of Ca2+ influx by neurosteroids in murine embryonic sensory neurones. Cell Calcium 40:383–391
764. Purves-Tyson TD, Arshi MS, Handelsman DJ, Cheng Y, Keast JR (2007) Androgen and estrogen receptor-mediated mechanisms of testosterone action in male rat pelvic autonomic ganglia. Neuroscience 148:92–104
765. Koenig HL, Schumacher M, Ferzaz B, Thi AN, Ressouches A, Guennoun R, Jung-Testas I, Robel P, Akwa Y, Baulieu EE (1995) Progesterone synthesis and myelin formation by Schwann cells. Science 268:1500–1503
766. Akwa Y, Schumacher M, Jung-Testas I, Baulieu EE (1993) Neurosteroids in rat sciatic nerves and Schwann cells. C R Acad Sci III 316:410–414
767. Morfin R, Young J, Corpechot C, Egestad B, Sjovall J, Baulieu EE (1992) Neurosteroids: pregnenolone in human sciatic nerves. Proc Natl Acad Sci USA 89:6790–6793
768. Zhu TS, Glaser M (2008) Regulatory role of cytochrome P450scc and pregnenolone in myelination by rat Schwann cells. Mol Cell Biochem 313:79–89
769. Desarnaud F, Do Thi AN, Brown AM, Lemke G, Suter U, Baulieu EE, Schumacher M (1998) Progesterone stimulates the activity of the promoters of peripheral myelin protein-22 and protein zero genes in Schwann cells. J Neurochem 71:1765–1768
770. Azcoitia I, Leonelli E, Magnaghi V, Veiga S, Garcia-Segura LM, Melcangi RC (2003) Progesterone and its derivatives dihydroprogesterone and tetrahydroprogesterone reduce myelin fiber morphological abnormalities and myelin fiber loss in the sciatic nerve of aged rats. Neurobiol Aging 24:853–860
771. Girard C, Liu S, Cadepond F, Adams D, Lacroix C, Verleye M, Gillardin JM, Baulieu EE, Schumacher M, Schweizer-Groyer G (2008) Etifoxine improves peripheral nerve regeneration and functional recovery. Proc Natl Acad Sci USA 105:20505–20510
772. Chavez-Delgado ME, Gomez-Pinedo U, Feria-Velasco A, Huerta-Viera M, Castaneda SC, Toral FA, Parducz A, Anda SL, Mora-Galindo J, Garcia-Estrada J (2005) Ultrastructural analysis of guided nerve regeneration using progesterone- and pregnenolone-loaded chitosan prostheses. J Biomed Mater Res B Appl Biomater 74:589–600

773. Schaeffer V, Meyer L, Patte-Mensah C, Eckert A, Mensah-Nyagan AG (2010) Sciatic nerve injury induces apoptosis of dorsal root ganglion satellite glial cells and selectively modifies neurosteroidogenesis in sensory neurons. Glia 58:169–180
774. Patte-Mensah C, Meyer L, Schaeffer V, Mensah-Nyagan AG (2010) Selective regulation of 3 alpha-hydroxysteroid oxido-reductase expression in dorsal root ganglion neurons: a possible mechanism to cope with peripheral nerve injury-induced chronic pain. Pain 150:522–534
775. Sereda MW, Meyer zu HG, Suter U, Uzma N, Nave KA (2003) Therapeutic administration of progesterone antagonist in a model of Charcot-Marie-Tooth disease (CMT-1A). Nat Med 9:1533–1537
776. Cermenati G, Giatti S, Cavaletti G, Bianchi R, Maschi O, Pesaresi M, Abbiati F, Volonterio A, Saez E, Caruso D, Melcangi RC, Mitro N (2010) Activation of the liver X receptor increases neuroactive steroid levels and protects from diabetes-induced peripheral neuropathy. J Neurosci 30:11896–11901
777. Leonelli E, Bianchi R, Cavaletti G, Caruso D, Crippa D, Garcia-Segura LM, Lauria G, Magnaghi V, Roglio I, Melcangi RC (2007) Progesterone and its derivatives are neuroprotective agents in experimental diabetic neuropathy: a multimodal analysis. Neuroscience 144:1293–1304
778. Sicard F, Ehrhart-Bornstein M, Corbeil D, Sperber S, Krug AW, Ziegler CG, Rettori V, McCann SM, Bornstein SR (2007) Age-dependent regulation of chromaffin cell proliferation by growth factors, dehydroepiandrosterone (DHEA), and DHEA sulfate. Proc Natl Acad Sci USA 104:2007–2012
779. Ziegler CG, Langbein H, Krug AW, Ludwig B, Eisenhofer G, Ehrhart-Bornstein M, Bornstein SR (2011) Direct effect of dehydroepiandrosterone sulfate (DHEAS) on PC-12 cell differentiation processes. Mol Cell Endocrinol 336:149–155
780. Ziegler CG, Sicard F, Lattke P, Bornstein SR, Ehrhart-Bornstein M, Krug AW (2008) Dehydroepiandrosterone induces a neuroendocrine phenotype in nerve growth factor-stimulated chromaffin pheochromocytoma PC12 cells. Endocrinology 149:320–328
781. Alexaki VI, Dermitzaki E, Charalampopoulos I, Kampa M, Nifli AP, Gravanis A, Margioris AN, Castanas E (2006) Neuronal differentiation of PC12 cells abolishes the expression of membrane androgen receptors. Exp Cell Res 312:2745–2756
782. Chadha HK, Hubscher CH (2008) Convergence of nociceptive information in the forebrain of female rats: reproductive organ response variations with stage of estrus. Exp Neurol 210:375–387
783. Qiao GF, Li BY, Lu YJ, Fu YL, Schild JH (2009) 17Beta-estradiol restores excitability of a sexually dimorphic subset of myelinated vagal afferents in ovariectomized rats. Am J Physiol Cell Physiol 297:C654–C664
784. Wiebe JP, Kavaliers M (1988) Analgesic effects of the putative FSH-suppressing gonadal steroid, 3 alpha-hydroxy-4-pregnen-20-one: possible modes of action. Brain Res 461:150–157
785. Kavaliers M, Wiebe JP (1987) Analgesic effects of the progesterone metabolite, 3 alpha-hydroxy-5 alpha-pregnan-20-one, and possible modes of action in mice. Brain Res 415:393–398
786. Frye CA, van Keuren KR, Rao PN, Erskine MS (1996) Analgesic effects of the neurosteroid 3 alpha-androstanediol. Brain Res 709:1–9
787. Kavaliers M, Perrot-Sinal TS, Desjardins DC, Cross-Mellor SK, Wiebe JP (2000) Antinociceptive effects of the neuroactive steroid, 3alpha-hydroxy-5alpha-pregnan-20-one and progesterone in the land snail, *Cepaea nemoralis*. Neuroscience 95:807–812
788. Patte-Mensah C, Kibaly C, Mensah-Nyagan AG (2005) Substance P inhibits progesterone conversion to neuroactive metabolites in spinal sensory circuit: a potential component of nociception. Proc Natl Acad Sci USA 102:9044–9049
789. Poisbeau P, Patte-Mensah C, Keller AF, Barrot M, Breton JD, Luis-Delgado OE, Freund-Mercier MJ, Mensah-Nyagan AG, Schlichter R (2005) Inflammatory pain upregulates spinal inhibition via endogenous neurosteroid production. J Neurosci 25:11768–11776
790. Charlet A, Lasbennes F, Darbon P, Poisbeau P (2008) Fast non-genomic effects of progesterone-derived neurosteroids on nociceptive thresholds and pain symptoms. Pain 139:603–609

791. Sasso O, Russo R, Vitiello S, Raso GM, D'Agostino G, Iacono A, Rana GL, Vallee M, Cuzzocrea S, Piazza PV, Meli R, Calignano A (2012) Implication of allopregnanolone in the antinociceptive effect of N-palmitoylethanolamide in acute or persistent pain. Pain 153:33–41
792. Wang X, Dyson MT, Jo Y, Stocco DM (2003) Inhibition of cyclooxygenase-2 activity enhances steroidogenesis and steroidogenic acute regulatory gene expression in MA-10 mouse Leydig cells. Endocrinology 144:3368–3375
793. Kawano T, Soga T, Chi H, Eguchi S, Yamazaki F, Kumagai N, Yokoyama M (2011) Role of the neurosteroid allopregnanolone in the hyperalgesic behavior induced by painful nerve injury in rats. J Anesth 25:942–945
794. Kawano T, Soga T, Chi H, Eguchi S, Yamazaki F, Yokoyama M (2011) The involvement of the neurosteroid allopregnanolone in the antihyperalgesic effect of paroxetine in a rat model of neuropathic pain. Neuroreport 22:984–988
795. Vriens J, Owsianik G, Hofmann T, Philipp SE, Stab J, Chen X, Benoit M, Xue F, Janssens A, Kerselaers S, Oberwinkler J, Vennekens R, Gudermann T, Nilius B, Voets T (2011) TRPM3 is a nociceptor channel involved in the detection of noxious heat. Neuron 70:482–494
796. Ueda H, Inoue M, Yoshida A, Mizuno K, Yamamoto H, Maruo J, Matsuno K, Mita S (2001) Metabotropic neurosteroid/sigma-receptor involved in stimulation of nociceptor endings of mice. J Pharmacol Exp Ther 298:703–710
797. Chen SC, Liu BC, Chen CW, Wu FS (2006) Intradermal pregnenolone sulfate attenuates capsaicin-induced nociception in rats. Biochem Biophys Res Commun 349:626–633
798. Yoon SY, Roh DH, Seo HS, Kang SY, Moon JY, Song S, Beitz AJ, Lee JH (2010) An increase in spinal dehydroepiandrosterone sulfate (DHEAS) enhances NMDA-induced pain via phosphorylation of the NR1 subunit in mice: involvement of the sigma-1 receptor. Neuropharmacology 59:460–467
799. Kibaly C, Meyer L, Patte-Mensah C, Mensah-Nyagan AG (2008) Biochemical and functional evidence for the control of pain mechanisms by dehydroepiandrosterone endogenously synthesized in the spinal cord. FASEB J 22:93–104
800. Riley JL III, Robinson ME, Wise EA, Myers CD, Fillingim RB (1998) Sex differences in the perception of noxious experimental stimuli: a meta-analysis. Pain 74:181–187
801. Riley JL III, Robinson ME, Wise EA, Price DD (1999) A meta-analytic review of pain perception across the menstrual cycle. Pain 81:225–235
802. Gaumond I, Arsenault P, Marchand S (2002) The role of sex hormones on formalin-induced nociceptive responses. Brain Res 958:139–145
803. Gaumond I, Arsenault P, Marchand S (2005) Specificity of female and male sex hormones on excitatory and inhibitory phases of formalin-induced nociceptive responses. Brain Res 1052:105–111
804. Aloisi AM, Ceccarelli I (2000) Role of gonadal hormones in formalin-induced pain responses of male rats: modulation by estradiol and naloxone administration. Neuroscience 95:559–566
805. Lu YC, Chen CW, Wang SY, Wu FS (2009) 17Beta-estradiol mediates the sex difference in capsaicin-induced nociception in rats. J Pharmacol Exp Ther 331:1104–1110
806. Spooner MF, Robichaud P, Carrier JC, Marchand S (2007) Endogenous pain modulation during the formalin test in estrogen receptor beta knockout mice. Neuroscience 150:675–680
807. Okuda K, Iwasaka H, Hagiwara S, Takeshima N, Takatani J, Uchino T, Noguchi T (2011) The antinociceptive effects of estradiol on adjuvant-induced hyperalgesia in rats involve activation of adrenergic and serotonergic systems. J Anesth 25:392–397
808. Dina OA, Aley KO, Isenberg W, Messing RO, Levine JD (2001) Sex hormones regulate the contribution of PKCepsilon and PKA signalling in inflammatory pain in the rat. Eur J Neurosci 13:2227–2233
809. Chaban VV, Mayer EA, Ennes HS, Micevych PE (2003) Estradiol inhibits atp-induced intracellular calcium concentration increase in dorsal root ganglia neurons. Neuroscience 118:941–948

810. Chaban VV, Micevych PE (2005) Estrogen receptor-alpha mediates estradiol attenuation of ATP-induced Ca2+ signaling in mouse dorsal root ganglion neurons. J Neurosci Res 81:31–37
811. Liu NJ, Gintzler AR (2000) Prolonged ovarian sex steroid treatment of male rats produces antinociception: identification of sex-based divergent analgesic mechanisms. Pain 85:273–281
812. Micevych PE, Rissman EF, Gustafsson JA, Sinchak K (2003) Estrogen receptor-alpha is required for estrogen-induced mu-opioid receptor internalization. J Neurosci Res 71: 802–810
813. Frodl T, O'Keane V (2012) How does the brain deal with cumulative stress? A review with focus on developmental stress, HPA axis function and hippocampal structure in humans. Neurobiol Dis (in press)
814. Goosens KA, Sapolsky RM (2007) Stress and glucocorticoid contributions to normal and pathological aging. In: Riddle D (ed) Brain aging: models, methods, and mechanisms. CRC Press, Boca Raton
815. Guazzo EP, Kirkpatrick PJ, Goodyer IM, Shiers HM, Herbert J (1996) Cortisol, dehydroepiandrosterone (DHEA), and DHEA sulfate in the cerebrospinal fluid of man: relation to blood levels and the effects of age. J Clin Endocrinol Metab 81:3951–3960
816. Geerling JC, Loewy AD (2009) Aldosterone in the brain. Am J Physiol Renal Physiol 297:F559–F576
817. Gomez-Sanchez EP, Ahmad N, Romero DG, Gomez-Sanchez CE (2005) Is aldosterone synthesized within the rat brain? Am J Physiol Endocrinol Metab 288:E342–E346
818. Yu Y, Wei SG, Zhang ZH, Gomez-Sanchez E, Weiss RM, Felder RB (2008) Does aldosterone upregulate the brain renin-angiotensin system in rats with heart failure? Hypertension 51:727–733
819. Lovelace M, Watson TG, Stephenson GL (2003) Steroid 21-hydroxylase expression in cultured rat astrocytes. Brain Res Bull 61:609–615
820. Stromstedt M, Waterman MR (1995) Messenger RNAs encoding steroidogenic enzymes are expressed in rodent brain. Brain Res Mol Brain Res 34:75–88
821. Beyenburg S, Watzka M, Clusmann H, Blumcke I, Bidlingmaier F, Elger CE, Stoffel-Wagner B (2001) Messenger RNA of steroid 21-hydroxylase (CYP21) is expressed in the human hippocampus. Neurosci Lett 308:111–114
822. Kishimoto W, Hiroi T, Shiraishi M, Osada M, Imaoka S, Kominami S, Igarashi T, Funae Y (2004) Cytochrome P450 2D catalyze steroid 21-hydroxylation in the brain. Endocrinology 145:699–705
823. Gomez-Sanchez CE, Zhou MY, Cozza EN, Morita H, Foecking MF, Gomez-Sanchez EP (1997) Aldosterone biosynthesis in the rat brain. Endocrinology 138:3369–3373
824. Gomez-Sanchez EP, Gomez-Sanchez CM, Plonczynski M, Gomez-Sanchez CE (2010) Aldosterone synthesis in the brain contributes to Dahl salt-sensitive rat hypertension. Exp Physiol 95:120–130
825. Huang BS, White RA, Ahmad M, Jeng AY, Leenen FH (2008) Central infusion of aldosterone synthase inhibitor prevents sympathetic hyperactivity and hypertension by central Na+in Wistar rats. Am J Physiol Regul Integr Comp Physiol 295:R166–R172
826. Huang BS, Ahmadi S, Ahmad M, White RA, Leenen FH (2010) Central neuronal activation and pressor responses induced by circulating ANG II: role of the brain aldosterone-"ouabain" pathway. Am J Physiol Heart Circ Physiol 299:H422–H430
827. Murrell JR, Randall JD, Rosoff J, Zhao JL, Jensen RV, Gullans SR, Haupert GT Jr (2005) Endogenous ouabain: upregulation of steroidogenic genes in hypertensive hypothalamus but not adrenal. Circulation 112:1301–1308
828. Bagrov AY, Shapiro JI, Fedorova OV (2009) Endogenous cardiotonic steroids: physiology, pharmacology, and novel therapeutic targets. Pharmacol Rev 61:9–38
829. Laredo J, Shah JR, Lu ZR, Hamilton BP, Hamlyn JM (1997) Angiotensin II stimulates secretion of endogenous ouabain from bovine adrenocortical cells via angiotensin type 2 receptors. Hypertension 29:401–407

830. Gomez-Sanchez EP, Samuel J, Vergara G, Ahmad N (2005) Effect of 3beta-hydroxysteroid dehydrogenase inhibition by trilostane on blood pressure in the Dahl salt-sensitive rat. Am J Physiol Regul Integr Comp Physiol 288:R389–R393
831. Leenen FH (2010) The central role of the brain aldosterone-"ouabain" pathway in salt-sensitive hypertension. Biochim Biophys Acta 1802:1132–1139
832. Gomez-Sanchez CE, Zhou MY, Cozza EN, Morita H, Eddleman FC, Gomez-Sanchez EP (1996) Corticosteroid synthesis in the central nervous system. Endocr Res 22:463–470
833. Prager EM, Brielmaier J, Bergstrom HC, McGuire J, Johnson LR (2010) Localization of mineralocorticoid receptors at mammalian synapses. PLoS One 5:e14344
834. Mihailidou AS (2006) Nongenomic actions of aldosterone: physiological or pathophysiological role? Steroids 71:277–280
835. Grossmann C, Gekle M (2009) New aspects of rapid aldosterone signaling. Mol Cell Endocrinol 308:53–62
836. Gros R, Ding Q, Sklar LA, Prossnitz EE, Arterburn JB, Chorazyczewski J, Feldman RD (2011) GPR30 expression is required for the mineralocorticoid receptor-independent rapid vascular effects of aldosterone. Hypertension 57:442–451
837. Walf AA, Rhodes ME, Meade JR, Harney JP, Frye CA (2007) Estradiol-induced conditioned place preference may require actions at estrogen receptors in the nucleus accumbens. Neuropsychopharmacology 32:522–530
838. Cagetti E, Pinna G, Guidotti A, Baicy K, Olsen RW (2004) Chronic intermittent ethanol (CIE) administration in rats decreases levels of neurosteroids in hippocampus, accompanied by altered behavioral responses to neurosteroids and memory function. Neuropharmacology 46:570–579
839. Charlier TD, Harada N, Balthazart J, Cornil CA (2011) Human and quail aromatase activity is rapidly and reversibly inhibited by phosphorylating conditions. Endocrinology 152:4199–4210
840. King SR, Bhangoo A, Stocco DM (2011) Functional and physiological consequences of StAR deficiency: role in lipoid congenital adrenal hyperplasia. Endocr Dev 20:47–53
841. Abdulhadi-Atwan M, Jean A, Chung WK, Meir K, Ben NZ, Stratigopoulos G, Oberfield SE, Fennoy I, Hirsch HJ, Bhangoo A, Ten S, Lerer I, Zangen DH (2007) Role of a founder c.201_202delCT mutation and new phenotypic features of congenital lipoid adrenal hyperplasia in Palestinians. J Clin Endocrinol Metab 92:4000–4008
842. Unger C, Eibl H, von Heyden HW, Kim DJ, Nagel GA (1986) Aminoglutethimide. Penetration of the blood brain barrier. Invest New Drugs 4:237–240
843. Lephart ED (1995) Age-related changes in brain and pituitary 5 alpha-reductase with finasteride (Proscar) treatment. Neurobiol Aging 16:647–650
844. Parepally JM, Mandula H, Smith QR (2006) Brain uptake of nonsteroidal anti-inflammatory drugs: ibuprofen, flurbiprofen, and indomethacin. Pharm Res 23:873–881
845. Akk G, Covey DF, Evers AS, Steinbach JH, Zorumski CF, Mennerick S (2007) Mechanisms of neurosteroid interactions with GABA(A) receptors. Pharmacol Ther 116:35–57
846. Cobos EJ, Entrena JM, Nieto FR, Cendan CM, del Pozo E (2008) Pharmacology and therapeutic potential of sigma(1) receptor ligands. Curr Neuropharmacol 6:344–366
847. Sakamoto H, Ukena K, Tsutsui K (2002) Dendritic spine formation in response to progesterone synthesized de novo in the developing Purkinje cell in rats. Neurosci Lett 322:111–115
848. McCarthy MM (2007) GABA receptors make teens resistant to input. Nat Neurosci 10:397–399
849. Liu A, Margaill I, Zhang S, Labombarda F, Coqueran B, Delespierre B, Liere P, Marchand-Leroux C, O'Malley BW, Lydon JP, De Nicola AF, Sitruk-Ware R, Mattern C, Plotkine M, Schumacher M, Guennoun R (2012) Progesterone receptors: a key for neuroprotection in experimental stroke. Endocrinology 153:3747–3757

Index

S.R. King, *Neurosteroids and the Nervous System*, SpringerBriefs in Neuroscience, DOI 10.1007/978-1-4614-5559-2, © Springer Science+Business Media, LLC 2013

Printed by Books on Demand, Germany